DASH Diet

The Essential Dash Diet Cookbook for Beginners – Delicious Dash Diet Recipes for Optimal Weight Loss and Healthy Living

Whitney Harris

Table of Contents

Introduction

Greetings aspiring dieter! Thank you so much for buying *The Essential Dash Diet Cookbook for Beginners*. I fully understand all the different diets available all being hyped up and the DASH diet being wedged in as one of them. But if you are too afraid to give up your favorite foods and make all sorts of complicated sacrifices just to become healthier, you made the right choice of considering the DASH diet!

The DASH Diet is unlike other diets as it has a core focus of lowering your blood pressure. Take it right from the meaning of **DASH** which stands for Dietary Approaches to Stop Hypertension. The best part? The rewards to sticking with the DASH diet not only comes in the form of regulating your blood pressure but also weight loss for the long-term! A win-win for sure!

After reading this book you should be able to know and make the following changes to whatever diet you currently have now:
- How to enjoy all of your favorite foods in moderation without exceeding the recommended daily salt consumption.
- How to choose the right ingredients to make some super delicious recipes that conform with the DASH diet standards.
- How to adjust your drinking lifestyle if you are into carbonated, alcoholic, and/or caffeine beverages.
- How to limit yourself from eating sweet and salty foods without leaving your taste buds wanting more.
- How exercising fits in to the DASH diet lifestyle.
- How the DASH diet can benefit those with diabetes.

The DASH diet is one of the easiest diets to adjust to and there are plenty of yummy recipes to get you started. So read on and enjoy as you embark on a quest to get rid of all the bad fats and excess sodium into your system and significantly reduce the chances of hypertension, heart attack and stroke.

Chapter 1: What is the DASH Diet?

The DASH diet is a dietary pattern where in new adopters will have to adjust the amount of servings of certain foods daily with the goal of controlling and preventing hypertension. By transitioning to this diet, you will be encouraged to eat more fruits, vegetables, low-fat meat and dairy food and whole grains.

Unlike the fad diets being hyped up on the Internet and full of debates, the DASH diet is actually being promoted by an agency of the U.S. Department of Health and Human Services. Hypertension affects over a billion people worldwide and around 50 million people in the US alone. It is a very serious condition as it paves the way to strokes, kidney diseases and heart attacks. This led the National Heart, Lung, and Blood Institute (NHLBI) to conduct various studies on adults aged 30 to 50 trying out different dietary plans. The results were compared to typical American diets and noticed great improvements to blood pressure and significantly reduced LDL cholesterol levels. So the DASH Diet is the end result of a US-based study and acknowledged as a healthy eating pattern by the U.S. Department of Agriculture in their 2010 edition of their Dietary Guidelines for Americans handbook. Further studies by the same institute have shown that the DASH diet is effective with weight loss when combined with an increase in physical activity. Basically, a bunch of trial and error with serving sizes on fat, sugar and other staple foods gave birth to the DASH diet.

The DASH Diet tackles hypertension directly as the foods the diet encourages you to eat are often rich in potassium. Potassium alone specializes in reducing the effects of sodium which in turn lessens the blood pressure. More modernized versions of the DASH Diet go as far as suggesting to reduce your sodium intake as well to less than 2,300 mg a day. Those who have higher salt sensitivity should consider taking in not more than 1,500 mg of sodium a day.

High blood pressure isn't just related to hypertension but also related to type 2 diabetes. The early stage of type 2 diabetes involves your body not properly responding to insulin. Having a high insulin resistance means that your body cannot properly metabolize glucose leading to high blood pressure. The direct way to reverse insulin resistance is to lose weight and the way to lose weight is to cut down on calories. How this all links to the DASH diet will surprise you!

DASH Diet Benefits for Healthy Living Reduce and Prevent Hypertension

Hypertension is within the DASH Diet's acronym, thus making hypertension a primary focus. The various studies on hypertension which led to the formation of the DASH diet fully proves its effectiveness in reducing hypertension.

Hypertension is the shortened term for high blood pressure. It occurs when blood forcibly pushes through your blood vessels at high rates. Blood pressure is measured through two figures – the systolic pressure and diastolic pressure. These two numbers (where the systolic pressure is found on top and the diastolic pressure is found on the bottom) determine the category your blood pressure level is in. Having a systolic pressure of 140 mmHg or higher and a diastolic pressure of 90 mmHg or higher is considered hypertension. There is also an entry-level category known as prehypertension where the systolic pressure falls between 120 – 139 mmHg and the diastolic pressure goes between 80 – 89 mmHg. Your pressure level is only considered normal if the systolic and diastolic pressure figures are 120 and 80 mmHg or less respectively. But even after your blood pressure levels are normal, sticking to the DASH Diet is still good for you in the long term and you should be far less likely to get high blood pressure again.

Fruits and vegetables which are high in calcium, magnesium, potassium and antioxidants are key to reducing blood pressure. That's exactly the kind of stuff the DASH Diet encourages you to eat.

Gradual Weight Loss

For a diet that isn't even centered on weight loss, losing weight is a pleasant aftereffect for keeping the sodium away. While your mileage may vary depending on your exercising habits and other physical activities, you can consider DASH Diet as a stress-free way to lose weight. Reading labels are still important since you have to make sure the sodium content is low but you don't have to stress yourself out so much with counting calories and the like.

Helps Control Diabetes

Losing weight as an aftereffect gives way to another key benefit which happens to be great for diabetic patients. Even those who do not have diabetes should keep this in mind because it isn't very obvious if you have insulin resistance and leaving that untreated could lead to type 2 diabetes. The best fruits and vegetables for DASH diet happen to be low on calories as well and should fill up your tummy quite nicely so you won't be tempted to snack on some salty temptations. Keep that up and you could reverse that insulin resistance!

Easy to Adapt

There are several diets that can be too intimidating because they require significant changes to your eating habits. In some cases, you might even experience a "flu" as your body tries to adapt to the lack of nutrients like carbohydrates. The DASH diet isn't very complicated and should be relatively easy to transition to from another diet. Like other diets, a gradual change in eating habits is the way to go but with DASH diet, there isn't so much to keep in mind. For instance, if you are used to having a couple of vegetables per day, add a serving or two to one of your meals. If you are not used to eating so many fruits, you can try drinking natural fruit juices to start things off. Slowly part ways with soda and start drinking skim milk. Take that big leap from white rice and white bread to brown rice and wheat bread.

Even vegans can go onboard with the DASH diet as there are plenty of fruits and vegetables and the usual plant-based meat alternatives that have the essential nutrients for DASH diet. Think nuts, dry beans, lentils and seeds.

There are not any serious shifts to what you have to eat. It is more of keeping track of your daily intake of sodium and this guide is here to help you plan your servings!

Health Goals

Like just about any other diet, the DASH diet isn't the kind of diet where you can expect some very noticeable results in a week or two. Therefore, to stay motivated, it is best to set some milestone health goals and log your progress. Consider the sample chart below.

	Week 1		Week 2		Week 3		Week 4	
	Current	Target	Current	Target	Current	Target	Current	Target
Weight								
Waist Size								
Blood Pressure								
Sugar Level								
Cholesterol Level								

Chapter 2: Foods Do & Don't

Eat Freely

One of the nicest things about the DASH Diet is that pretty much all of the foods you know are healthy are fair game! What you are looking for is foods rich in calcium, fiber, magnesium and potassium. Use the list below as a guide:

- **Grains (6 to 8 servings per day)** – Barley, Brown Rice, Oatmeal, Wheat Bread, Wheat Pasta, Wheat Tortillas
- **Meats (Not more than 6 servings per day)** – Eggs, Lean Beef, Lean Chicken, Lean Pork, Turkey
- **Seafoods (Not more than 6 servings per day)** – Fish, Salmon, Shrimp
- **Fruits (4 to 5 servings per day)** – Apples, Bananas, Blackberries, Blueberries, Cherries, Grapes, Lemons, Limes, Mangoes, Peaches, Pears, Pineapples, Raspberries, Strawberries
- **Vegetables (4 to 5 servings per day)** – Artichokes, Bell Peppers, Broccoli, Brussels Sprout, Cabbage, Carrots, Cauliflower, Corn, Green Beans, Mushrooms, Lettuce, Onions, Squash

Eat Occasionally

On the DASH Diet, you should limit the following foods and ingredients to not more than 3 servings per day.

- **Vegetable Oils (2 to 3 servings per day)** – Canola Oil, Corn Oil, Olive Oil, Safflower Oil
- **Condiments (2 to 3 servings per day)** – Mayonnaise, Salad Dressing
- **Dairy (2 to 3 servings per day)** – Greek Yogurt, Low-fat Cheese, Low-fat Milk, Margarine, Skim Milk, Sour Cream
- **Nuts, Legumes and Seeds (4 to 5 servings per week)** – Almonds, Cashews, Flax Seeds, Hazelnuts, Kidney Beans, Lentils, Pecans, Pumpkin Seeds, Split Peas, Sunflower Seeds, Unsalted Peanuts, Walnuts
- **Red Meats (1 to 2 servings per week)**

Eat Rarely

Perhaps the biggest sacrifice you have to make when switching to the DASH Diet is giving up those salty foods. Also take note of the foods you should avoid eating as much as possible.

- **Sweets** – Beverages, Candies, Jams, Jellies, Sugars, Sweet Yogurt
- **Saturated Fats** – Bacon, Cholesterol, Coconuts, Fatty meats, Full-fat dairy
- **Sodium** – Canned Fruit, Canned Vegetables, Cookies, Chips, Frozen Vegetables, Gravy, Pizza, Salted Nuts

Chapter 3: DASH Diet Frequently Asked Questions (FAQ)

Should I eliminate salt completely from my diet?

No. Eating too little salt as it increases the risk of fluid retention and heart disease. Try to keep your sodium intake below 1,500 mg per day.

What about alcohol and caffeine?

Heavy beer drinkers and coffee lovers should keep things a bit moderate. Although they do not directly violate anything the DASH Diet is up for, drinking too much alcohol or coffee can increase blood pressure. A single drink per day for women or two for men shouldn't cause any harm. Just don't exceed that.

Is exercising necessary?

Not completely needed but exercising should be part of your daily regimen to make your body feel good. A simple 15-minute walk or jog everyday can make a world of difference in lowering blood pressure. Also try to bike or swim every now and then. If you lack the time, don't forget that spending an hour or so on household chores can count as exercising as well.

Is grass-fed beef big deal?

Yes it is even despite the higher price because grain feeding is known to destroy the omega-3 fatty acids normally found in beef. On top of that, grain-fed meat is a high content of omega-6 as well as saturated fat which contribute to high blood pressure, inflammation and obesity.

Should I choose DASH Diet to achieve my weight loss goals?

Maybe. DASH Diet could be a good choice to lose weight if you don't want to restrict yourself heavily like other diets which may require you to avoid high-carb foods for example. But try to temper your expectations as the main purpose of the DASH Diet is to regulate your blood pressure without requiring medications. You can view weight loss as one of the "side effects" to sticking with this diet but results can vary. If you like the freedom in choosing what you want to eat, be sure to spend a good amount of time exercising too to make the weight loss benefits more noticeable.

Will Calcium, Potassium and Magnesium Supplements Help?

Nothing beats the real thing. The best way to stick to the DASH Diet is to take foods rich in minerals and avoid foods that will raise your cholesterol levels. Besides, none of the supplements are particularly effective in lowering cholesterol.

Chapter 4: Breakfast & Smoothie Recipes

No matter what diet you stick to, breakfast remains as the most important meal of the day and the DASH Diet is no exception to that. And cutting down on the salt and sweet stuff doesn't mean that you have to make your meals boring and flavorless too. Check out these awesome DASH Diet-compliant breakfast and smoothie recipes to start your day bursting with energy and liveliness!

Breakfast Bread Pudding

Cooking Time: 1 hour 15 minutes
Serves: 4

Ingredients:

- 2 1/4 cups low fat 1% or fat free milk
- 6 eggs
- 3 tablespoons brown sugar
- 1 teaspoon vanilla extract
- 3/4 teaspoon ground cinnamon
- 1/8 teaspoon salt
- 6-7 slices whole wheat bread, cubed
- 3/4 cup peeled and diced apple
- 6 tablespoons raisins
- 3 teaspoons powdered sugar (optional)

Directions:

1. In large bowl whisk together milk, eggs, brown sugar, vanilla, cinnamon and salt.

2. Add bread, apple, and raisins and mix well.

3. Transfer into a greased baking dish. Cover the dish with foil and bake in a preheated oven at 350 degrees for about 40-45 minutes.

4. Take out the foil and bake until the crust is golden brown.

5. Remove from the oven and cool for at least 15 – 20 minutes.

6. Sprinkle powdered sugar and serve warm.

Breakfast Apple and Raisin Oatmeal

Cooking Time: 6 hours
Serves: 2

Ingredients:

- 2 cups milk low-fat milk, soy or almond milk will also work, you can also use water
- 2 tablespoons honey or stevia
- 1 tablespoon light butter
- 1/2 teaspoon ground cinnamon
- 2 drops vanilla essence
- 1 cup old fashioned oats
- 1 cup apple chopped
- 1/4 cup raisins

Directions:

1. Use a Crock Pot Liner or spray the inside of a 5-quart slow cooker with non-stick cooking spray

2. Combine all ingredients to the slow cooker and stir well to combine.

3. Cover and cook on LOW overnight, ideally no more than 6 hours or it will dry out.

4. Stir well in the morning before serving.

Egg Toast with Avocado Spread

Cooking Time: 5 minutes
Serves: 1

Ingredients:

- 2 eggs
- 2 slices of whole wheat bread
- 1 avocado
- 1 teaspoon lime juice
- ground black pepper

Directions:

1. Toast wheat bread and cook eggs to your liking.

2. Peel and mash the avocado so you can later spread it on your bread. Add pepper and lime juice.

3. Evenly spread the mashed avocado on each bread slice.

4. Add the egg on top and serve immediately.

Breakfast Green Smoothie

Cooking Time: 5 minutes
Serves: 1

Ingredients:

- 2 medium bananas, peeled, chopped
- 2 cups baby spinach, packed
- 1 cup fat-free milk
- 1/2 cup whole oats
- 1 1/2 cups frozen mango
- 1/2 cup plain nonfat yogurt
- 1 teaspoon vanilla
- Ice cubes

Directions:

1. Add milk, yogurt and oats to the blender and blend for about 12-15 seconds.

2. Add the rest of the Ingredients and blend until smooth.

3. Pour into tall glasses and serve immediately.

Non-fat Strawberry Banana Smoothie

Cooking Time: 5 minutes
Serves: 1

Ingredients:

- 6 chopped pieces of frozen strawberries
- 2 sliced pieces of fresh strawberries
- 1 banana
- 1/2 cup soy milk or non-fat milk
- 1 cup nonfat vanilla yogurt

Directions:

1. Mix all ingredients (except the fresh strawberry slices) together in a blender until mixture is smooth.

2. Pour the smoothie in a glass or pitcher

3. Add fresh strawberries as garnish.

Cinnamon Breakfast Quinoa

Cooking Time: 1 hour
Serves: 2

Ingredients:

- 1/2 cup quinoa, rinsed, drained
- 3/4 cup water
- 1/8 teaspoon salt
- 1 stick cinnamon
- Honey or maple syrup to serve
- 2 tablespoons walnuts, chopped
- Milk to serve

Directions:

1. Place a heavy bottomed saucepan over a medium heat. Add quinoa, water, salt and cinnamon. Mix well and bring to a boil.

2. Lower the heat, cover and simmer until all the water is absorbed.

3. Remove from the heat. Do not uncover for 5 minutes.

4. Fluff the cooked quinoa with a fork and discard the cinnamon stick.

5. Divide the quinoa into individual serving bowls. Pour milk over the quinoa.

6. Drizzle honey over the quinoa, sprinkle walnuts on top and serve.

Creamy Fruity French Toast

Cooking Time: 10 minutes
Serves: 8

Ingredients:

- 8 bread slices
- 2 egg whites
- 1 egg, whisked
- 3/4 cup low-fat milk
- 1/2 cup low-fat cream cheese
- 2 tablespoons strawberries, chopped
- 1/2 cup strawberry fruit, spreadable
- 1/2 teaspoon vanilla extract
- A pinch of apple pie spice
- Cooking spray

Directions:

1. Combine the chopped strawberries and cream cheese in a bowl and mix them well.

2. Form a horizontal pocket for each French toast slice and the strawberry cream cheese mix in it.

3. Get another bowl and combine the egg white with the whisked egg, milk, and vanilla extract. Stir well while adding a pinch of apple pie spice.

4. Soak each bread slice in the eggs mixture.

5. Turn up your stove to medium high heat and place a pan sprayed with cooking spray on it. Add the bread slices and cook one side for 2 minutes. Flip them over and cook the other side for 2 minutes.

6. Transfer the spreadable fruit to a small pan and put it on your stove on medium heat.

7. Spread the creamy fruit over each of the French toast slices and serve.

Sunday Breakfast Apple Cinnamon Crisp

Cooking Time: 1 hour 10 minutes
Serves: 4

Ingredients:

- 1 Cup oatmeal
- 1 cup brown sugar
- 2 tablespoon of all-purpose flour
- 1 tablespoon granulated sugar
- 1 stick of butter
- 1 teaspoon cinnamon
- 3 lbs. Granny Smith Apples

Directions:

1. Peel and core the apples, slice thinly.

2. Mix granulated sugar with flour and add the apples. Toss to coat.

3. Put apples into the bottom of a 5-6 quart crock pot.

4. Combine brown sugar with oatmeal and butter. Mix until mixture is crumbly.

5. Sprinkle oatmeal mixture on top of the apples.

6. Cook apples fully on high heat.

Whole Wheat Sweet Potato Cakes

Cooking Time: 10 minutes
Serves: 4

Ingredients:

- 4 cups sweet potato, peeled and grated
- 1/4 cup whole wheat flour
- 1 egg, whisked
- 1 teaspoon onion, minced
- A pinch of salt
- A pinch of nutmeg
- Black pepper to the taste
- Cooking spray

Directions:

1. Get a bowl and add the grated sweet potatoes, wheat flour, onion, egg, nutmeg, salt and pepper. Stir all mentioned ingredients well.

2. Prepare a skillet and spray it with cooking spray. Place skillet on your stove over medium high heat.

3. Shape the cakes out of the mixture from the bowl and add them to pan. Cook each side for 4 minutes and transfer them to paper towels so you can drain any excess grease.

4. Divide the cakes and serve.

Whole Wheat-Oat Pancakes with Blueberry Compote

Cooking Time: 30 minutes
Serves: 2

Ingredients:

- 2/3 cup oats
- 1/2 cup all-purpose flour
- 1/2 cup whole wheat flour
- 1 tablespoon baking powder
- 1/4 teaspoon kosher salt
- 1 cup non-fat milk
- 1/4 cup egg substitute
- 1 1/2 tablespoons vegetable oil
- Cooking spray
- 1 tablespoon powdered sugar
- 3/4 cup low calorie maple syrup

Directions:

1. Ground the oats using a food processor or grinder.

2. Combine the oats, baking powder, all-purpose flour, whole wheat flour, and kosher salt in a bowl.

3. Add in the milk, egg substitute, and oil to the oats mixture you made earlier. Stir the batter until it moistens.

4. Spray your cooking spray on a skillet and pour 1/4 cup batter on it. You may also use a hot griddle.

5. Wait for the edges to cook and look for signs of bubbling on the top. Then flip the pancake and cook that side.

6. Transfer the pancakes to a plate forming a nice stack.

7. Sprinkle with icing sugar and add maple syrup to serve.

Detox Smoothie

Cooking Time: 5 minutes
Serves: 1

Ingredients:

- 2 green apples, juiced
- 2 green apples, peeled, cored, chopped
- 1 tablespoon fresh ginger, minced
- Juice of 2 lime
- 1 cup wild arugula
- Ice cubes

Directions:

1. Add all the Ingredients to a blender and blend until smooth.

2. Pour into tall glasses and serve immediately.

Fresh Fruit Crunch

Cooking Time: 15 minutes
Serves: 2

Ingredients:

- 2 cups mixed fresh fruits (mixture of any of the following fruits: orange, pineapple, apple, pear, peach, seedless grapes, grapefruit, kiwi fruit)
- 1 cup low fat vanilla yogurt
- 1 tablespoon honey
- 1/4 cup low fat granola
- 2 tablespoons coconut, toasted

Directions:

1. Divide the fruits into 2 tall glasses.

2. Divide the yogurt and spoon over the fruits.

3. Drizzle honey on top.

4. Top with granola and coconut. Refrigerate until chilled.

5. Serve cold.

Special Egg Scramble

Cooking Time: 10 minutes
Serves: 4

Ingredients:

- 3 eggs
- 1 1/2 cups eggs replacement
- 1/4 cup fat free milk
- 1/4 cup green bell pepper, chopped
- 3/4 cup tomato, chopped
- 1/4 cup green onions, chopped
- A pinch of salt
- A pinch of black pepper
- A splash of hot pepper
- Cooking spray

Directions:

1. Spray some cooking spray on a pan and heat it up to medium high heat. Add all chopped ingredients to the pan and sauté for 5 minutes. Transfer to a bowl afterwards.

2. Get another bowl, combine the rest of the ingredients and whisk them well.

3. Heat up the same pan you used earlier over medium high heat and add the eggs mixture. Keep cooking and stirring for a few minutes.

4. Serve with the sautéed veggies on top.

Zucchini-Lemon Muffins

Cooking Time: 30 minutes
Serves: 12

Ingredients:

- 2 cups all-purpose flour
- 1/2 cup sugar
- 1 tablespoon baking powder
- 1/4 teaspoon salt
- 1/4 teaspoon ground nutmeg
- 1 cup coarsely shredded zucchini
- 3/4 cup skim milk
- 3 tablespoons vegetable oil
- 1 egg
- Cooking spray

Directions:

1. Mix the first 6 ingredients in a bowl.

2. Create a well in the center of dry ingredients.

3. In a separate bowl, mix the zucchini, milk, oil, and egg, stir well.

4. Pour into the well of the dry ingredients and stir until just moistened.

5. Divide batter evenly among 12 muffin cups coated with cooking spray.

6. Bake at 400 F for 20 minutes or until golden.

7. Remove from the pan immediately and let the muffins cool on a wire rack.

Banana Cantaloupe Smoothie

Cooking Time: 5 minutes
Serves: 2

Ingredients:

- 1 banana, frozen and sliced
- 2-1/2 cups cantaloupe, peeled, frozen and cubed (1-inch)
- 5 ounces carton vanilla non-fat Greek yogurt
- 1/2 cup non-fat milk
- 1 teaspoon honey
- 1/2 cup ice

Directions:

1. Blend all ingredients except the cantaloupe with a blender until smooth.

2. Add the cubed cantaloupe pieces and blend until smooth.

3. Serve right away.

Cinnamon Pumpkin Waffles

Cooking Time: 30 minutes
Serves: 4

Ingredients:

- 1 Cup Pumpkin Puree
- 1 Cup White Whole-Wheat Flour
- 1/2 Cup Unbleached All-Purpose Flour
- ⅓ Cup Brown Sugar
- 11/2 Cups Low-Fat Milk
- 1 egg white
- 1 tablespoon canola oil
- 1 tablespoon sodium-free baking powder
- 1 tablespoon pure vanilla extract
- 2 teaspoons ground cinnamon
- 1/4 teaspoon ground allspice
- 1/4 teaspoon ground ginger

Directions:

1. Spray the waffle iron lightly with oil and preheat.

1. Combine all ingredients into a large mixing bowl.

2. Whisk until smooth.

3. Ladle batter onto the center of the hot waffle iron.

4. Close waffle iron and bake until golden brown, roughly 4–5 minutes.

5. Remove the cooked waffle from iron and repeat process with remaining batter, re-oiling waffle iron as necessary.

6. Serve immediately.

Summer Fruit Smoothie

Cooking Time: 5 minutes
Serves: 1

Ingredients:

- 1 cup fresh pineapple
- 1/2 cup cantaloupe or other melon
- 1 cup fresh strawberries
- Juice of 2 oranges
- 1 cup water
- 1 tablespoon honey

Directions:

1. Remove rind from pineapple and melon and cut into chunks. Hull the strawberries. Set fruit aside in the refrigerator.

2. Place all the ingredients into a blender and puree until smooth.

3. Serve cold.

Strawberry-Orange Low-fat Yogurt Milkshake

Preparation Time: 5 minutes
Serves: 2

Ingredients:

- 20 ounces strawberries, frozen and thawed
- 1 cup orange juice
- 6 ounces low-fat yogurt
- 1 cup low fat milk

Directions:

1. Blend all ingredients in a blender until smooth.

2. Fill up two glasses with the processed milkshake and serve.

Slim-down Smoothie

Cooking Time: 5 minutes
Serves: 1

Ingredients:

- 1 medium avocado
- 1 cup frozen blueberries (fresh or frozen)
- 1 tablespoon coconut oil
- 1/2 teaspoon ground cinnamon
- 2 tablespoons chia seeds
- 1 tablespoon honey (maple syrup or stevia for alternatives)
- 2 cups water
- Ice cubes

Directions:

1. Combine all of the listed ingredients and put them in your blender.

2. Blend until the mixture is smooth.

3. Pour mixture into a tall glass or pitcher. Serve immediately.

Cranberry Green Tea Smoothie

Cooking Time: 5 minutes
Serves: 1

Ingredients:

- 1/2 cup blueberries
- 1 cup cranberries
- 1 cup blackberries
- 10 whole strawberries
- 2 ripe bananas
- 1 cup plain soy milk
- 1 cup green tea, room temperature
- 4 tablespoons honey or packed light brown sugar
- Ice cubes

Directions:

1. Add all the Ingredients to a blender and blend until smooth.

2. Pour into tall glasses and serve with ice immediately.

Oat & Honey Porridge

Cooking Time: 15 minutes
Serves: 4

Ingredients:

- 1/2 cup rolled oats
- 3/4 cup quinoa
- 1/4 cup honey
- 3 cups boiling water

Directions:

1. Boil water in a saucepan and add the quinoa and rolled outs. Stir thoroughly.

2. Place saucepan over high heat and wait for porridge to boil.

3. Immediately turn the heat down to low. Cover and simmer for 10 minutes. Stir occasionally.

4. Remove from stove and add the honey. Stir until honey blended in well and serve immediately.

Herbed Wild Mushroom Oatmeal

Cooking Time: 1 hour
Serves: 2

Ingredients:

- 1 cup old fashioned oats
- 6 ounces mushrooms, sliced
- 2 stalks scallions, thinly sliced (keep the white and green parts separate)
- 1 teaspoon extra virgin olive oil
- 1 teaspoon lemon juice
- 2 cups water
- 1/2 teaspoon pepper powder or to taste
- 1 teaspoon fresh rosemary, chopped
- 2 eggs
- 1/2 teaspoon sea salt
- Cooking spray

Directions:

1. Place a skillet over medium high heat. Add oil. When the oil is hot, sauté mushrooms and the white part of the scallions.

2. Sauté until the mushrooms are cooked.

3. Add lemon juice and pepper.

4. Add water, rosemary and salt and bring to a boil. Add oats and the scallion greens and cook until the oats are cooked. Stir occasionally.

5. Meanwhile, place a nonstick pan over medium heat and spray with cooking spray.
6. Fry the eggs - sunny side up.

7. To serve: Place half the oat mixture over each serving plate. Sprinkle with cheese. Top with an egg and serve.

Layered Mango Green Smoothie

Cooking Time: 10 minutes
Serves: 4

Ingredients:

- 2 1/2 cups frozen mango pieces
- 2 cups frozen strawberries
- 2cups fresh spinach
- 1 cup water
- 1 cup almond milk plus extra
- Honey, stevia or agave to taste
- Ice cubes

Directions:

1. Add mango, water and milk to the blender and blend until smooth. If it is too thick add more water or milk. If it is too watery, then add more mangoes.

2. Divide and pour into 4 glasses.

3. Rinse the blender and add strawberries. Blend until smooth. Add more milk or water if necessary. Add ice, banana and sweetener and blend until smooth.

4. Pour or spoon over the mango layer in the glasses.

5. Clean the blender and add spinach, ice and a little milk and blend until smooth.
6. Pour or spoon over the strawberry layer.

7. Serve immediately.

Special Tofu Scramble

Cooking Time: 10 minutes
Serves: 4

Ingredients:

- 18 ounces packaged firm tofu
- 2 poblano chili peppers, chopped
- 2 garlic cloves, minced
- 1/2 cup onion, chopped
- 1 tablespoon olive oil
- 1 teaspoon chili powder
- 1/2 teaspoon cumin, ground
- 1/2 teaspoon oregano, dried
- A pinch of sea salt
- 2 tomatoes, peeled and chopped
- 1 tablespoon cilantro, chopped
- 1 tablespoon lime juice

Directions:

1. Take out the packaged tofu and drain it well. Pat the tofu dry and crumble it.

2. Pour the olive oil on a pan and heat it up over medium high heat. Add the garlic, onion and chili peppers and cook and stir for 4 minutes.

3. Add the oregano, cumin, chili powder and salt and continue cooking and stirring for a minute more.

4. Reduce the heat and add the crumbled tofu while still stirring for 5 more minutes.

5. Add the cilantro, lime juice and tomatoes. Stir well one last time and serve.

Irish Brown Soda Bread

Cooking Time: 1 hour
Serves: 4

Ingredients:

- 4 cups low fat buttermilk
- 4 cups whole wheat flour
- 3 cups all-purpose flour
- Extra all-purpose flour for dusting
- 1 cup wheat germ
- 2 eggs, beaten
- 4 teaspoons baking soda
- 1/2 teaspoon table salt

Directions:

1. Combine the dry Ingredients in a bowl.

2. Add eggs and low fat buttermilk on to the mixture and mix until it forms a dough.

3. Prepare a Transfer the dough onto a floured work area and knead gently forming a large round shape.

4. Place the kneaded bread on a nonstick baking sheet and make two half-inch deep cuts along the center forming an "X".

5. Preheat your oven at 400°F and put the bread there. Keep checking for signs of the bread splitting apart from the cuts you made and turn off the oven afterwards (takes about 25-30 minutes)

6. Remove the bread and put on a wire rack to cool.

7. Slice and serve.

Whole Wheat Walnut Pancakes

Cooking Time: 30 minutes
Serves: 2

Ingredients:

- 6 tablespoons all-purpose flour
- 6 tablespoons whole wheat flour
- 1/4 cup chopped walnuts
- 3/4 teaspoon baking powder
- 1/4 teaspoon salt
- 1/4 cup 2% milk
- 1 egg
- 1 1/2 tablespoons butter substitute spread, divided

Directions:

1. Mix together all the dry Ingredients in a large bowl.

2. Add milk, eggs and half of the melted butter substitute into a bowl and whisk well.

3. Pour the milk mixture into the flour mixture and whisk well until combined. Set aside.

4. Place a nonstick pan over medium heat. Add a little of the remaining butter substitute.

5. Pour about 1/4 cup batter on the pan. Swirl a bit to spread the pancake. When the bottom side is golden brown, flip and cook the other side until golden brown.

6. Repeat step 4 with the remaining batter.

Banana Peanut Butter Shake

Preparation Time: 5 minutes
Serves: 3

Ingredients:

- 1 frozen banana, sliced
- 1 tablespoon peanut butter
- 1/4 cup fat-free Greek yogurt
- 1 cup fat-free milk
- 1 cup ice

Directions:

1. Combine all of the ingredients together and pour them to your blender

2. Process the mixture until it smoothens.

3. Serve immediately.

Banana Oat Pancakes With Spiced Syrup

Cooking Time: 30 minutes
Serves: 2

Ingredients:

- 1/2 cup of maple syrup
- 1/2 stick of cinnamon
- 3 cloves
- 1/2 rolled oats
- 1 cup of water
- 2 tablespoons of brown sugar
- 2 tablespoons of canola oil
- 1/2 cup of whole-wheat flour

- 1 1/2 teaspoons of baking powder
- 1/4 teaspoon of salt
- Half cup of plain white flour
- 1/4 teaspoon of ground cinnamon
- 1 Peeled & mashed banana
- 1 Lightly beaten egg
- 1/4 Cup of fat-free yogurt (plain)

Directions:

1. Combine oats and water in a bowl, microwave it on high until oats are creamy.

2. Stir in brown sugar and canola oil, let the mixture cool to room temperature.

3. In a separate bowl, combine both flours, baking soda & powder, ground cinnamon and salt.

4. Mix in milk, yogurt and banana to the oats and stir until well blended. Beat in the egg & add flour to oats.

5. Heat a non-stick pan on medium heat and pour batter into it.

6. When a lot of bubbles rise to the top, flip the pancake and continue to cook until lightly browned on both sides.

7. Drizzle with syrup and serve warm.

DASH Oatmeal Special

Cooking Time: 5minutes
Serves: 4

Ingredients:

- 3 cups rolled oats
- 3 cups whole-grain cereal
- 4 cups water
- 1 cup dried dates, chopped
- 1 cup dried apples, chopped
- 1 cup walnuts, chopped

- 2 tablespoons cinnamon, ground
- 1 cup coconut sugar
- 1 tablespoon ginger, ground
- 1 teaspoon cloves, ground
- 1 teaspoon turmeric ground

Directions:

1. Boil water and prepare 4 separate bowls.

2. Distribute all ingredients among the bowls.

3. Add the boiling water to each the bowls and stir. Cover the bowls afterwards and set them aside for 10 minutes.

4. Serve immediately.

Spicy Strawberry Smoothie

Cooking Time: 5 minutes
Serves: 2

Ingredients:

- 1 pound strawberries
- 1 medium cucumber, peeled, roughly chopped
- 2 small Roma tomatoes
- 1 cup light coconut milk
- Juice of 2 limes
- 1 cup ice
- 1/2 teaspoon ground cayenne pepper or to taste
- Honey to taste
- A pinch of salt
- Lime slices to garnish
- Strawberries to garnish

Directions:

1. Add all the ingredients to a blender and blend until smooth.

2. Pour into tall glasses. Sprinkle some cayenne pepper. Garnish with strawberries and lime slices. Serve immediately.

Chapter 5: DASH Diet Main Course Recipes

The main course meals serve as the star of any dining table. When you want to eat them is entirely up to you may it be a quick lunch or that exotic dinner for that special occasion. Just remember to go light with the salt!

Grilled Chicken Salad

Cooking Time: 25 minutes
Serves: 2

Ingredients:

- 8 ounces grilled chicken breast, chopped into bite sized pieces
- 2 radishes, thinly sliced
- 1/2 cup pea shoots
- 1 pear, cored, peeled, sliced
- 2 cups arugula
- For dressing:
- 1 cup balsamic vinegar

- 1/2 cup dried figs
- 1/2 cup olive oil
- 4 tablespoons fresh basil, chopped
- 2 teaspoons lemon zest
- Salt to taste
- Pepper powder to taste

Directions:

1. To make dressing: Add figs, vinegar and oil to a blender and blend until smooth.

2. Transfer into a bowl. Add basil, lemon zest, salt and pepper. Mix well and set aside.

3. Place radish in a large salad bowl. Layer with pears, followed by pea shoots and arugula. Cover and refrigerate until use.

4. To serve: Remove salad from refrigerate. Place chicken over it. Drizzle the dressing over it and serve.

Seafood Chowder

Cooking Time: 45 minutes
Serves: 2

Ingredients:

- 2 small potatoes, cubed
- 1 carrot, cut into 1/4 inch thick slices
- 1 small onion, chopped
- 1/2 cup clam juice
- 1/2 cup water
- 1/2 tablespoon butter
- 1/8 teaspoon salt
- 1/8 teaspoon pepper
- 1/2 pound lean fish (cod, halibut or salmon) cut in small pieces
- 8 ounce canned clams, don't drain the water
- 6 ounce canned skim milk
- 1 tablespoon fresh chives, chopped
- 1/2 teaspoon paprika or to taste

Directions:

1. Heat a saucepan over medium heat. Add potatoes, carrots, onion, clam juice, water, butter, salt and pepper. Mix well and bring to a boil.

2. Lower heat, cover the saucepan, and simmer until the vegetables are cooked.

3. Add fish and clams. Raise the heat to high and bring to a boil.

4. Lower the heat and simmer for a few minutes until the fish flakes when pricked with a fork.

5. Add milk, chives and paprika. Bring to a boil.

6. Serve hot in soup bowls.

Roasted Garlic & Tomato Soup

Cooking Time: 1 hour 10 minutes
Serves: 4

Ingredients:

- 10 cups hot vegetable stock
- 2 1/2 pounds ripe tomatoes, quartered
- 8-9 garlic cloves, peeled
- 4 red onions, peeled and cut into wedges
- 2 red peppers, cubed and deseeded
- 1 tablespoon Worcestershire sauce
- 1 tablespoon balsamic vinegar
- 1/2 teaspoon kosher salt
- 1 tablespoon fresh basil
- Ground black pepper to taste
- Cooking spray

Directions:

1. Spray cooking spray on an ovenproof pan and place the tomatoes, red pepper, onions and garlic on top of it. Sprinkle with salt and black pepper.

2. Preheat your oven to 400°F and place the pan in it. Wait 45 minutes for the vegetables to roast until its edges are slightly charred and soft.

3. Remove the vegetables from the oven and let them cool.

4. Transfer the vegetables to your blender. Add the vegetable stock, Worcestershire sauce and vinegar. Blend until the mixture smoothens.

5. Transfer the mixture to a saucepan and heat it up.

6. Remove from heat and cool completely. Refrigerate immediately afterwards.

7. When ready to serve chilled, garnish with fresh basil.

Autumn Pork Chops

Cooking Time: 20 minutes
Serves: 4

Ingredients:

- 4 pork loin chops, bone-in
- 2 cups apple, cut into wedges
- 1/2 cup low sodium chicken stock
- 1/2 teaspoon white flour
- 2-1/2 teaspoons canola oil
- 1-1/2 cups pearl onions
- 1 teaspoon cider vinegar
- A pinch of sea salt
- Black pepper to the taste
- 2 teaspoons thyme, chopped

Directions:

1. Pour 1 teaspoon of oil into a hot pan over medium high heat.

2. Pat dry the pearl onions and add them to the pan. Cook for 2 minutes.

3. Add the apples and stir. Transfer pan to the oven and bake at 400°F for 10 minutes.

4. Take the pan out of the oven and add thyme, salt and pepper. Stir gently.

5. Prepare another pan with 1 and 1/2 teaspoon canola oil over medium high heat. Season the pork with salt and pepper.

6. Cook the pork for 3 minutes and then flip the pork and cook the other side for another 3 minutes. Remove from pan afterwards and keep warm.

7. Put chicken stock in a pot, add flour and stir well. Bring mixture to a boil and cook for a minute.

8. Stir in the apple cider vinegar and turn off the heat.

9. Serve the pork chops, apple and onion mix and sauce together.

Turkey Soup with Egg Noodles and Veggies

Cooking Time: 1 hour 5 minutes
Serves: 3

Ingredients:

- 4 ounces turkey breast meat, chopped or shredded into bite-sized pieces
- 4 cups turkey stock
- 8 ounces egg noodles
- 6 cups water
- 1 cup celery, diced
- 1 cup carrots, diced
- 1 medium onion, diced
- 1 teaspoon garlic, minced
- 1 teaspoon of salt (or to taste)
- 1 teaspoon ground black pepper

Directions:

1. Set the noodles aside and add the rest of the ingredients to large pot.

2. Boil the pot over high heat.

3. Turn down the heat, cover and simmer until vegetables are nearly cooked.

4. Uncover the pot and add the egg noodles. Simmer until they are soft.

5. Serve hot.

Ground Turkey Meatloaf

Cooking Time: 1 hour 30 minutes
Serves: 4

Ingredients:

- 3 pounds ground turkey
- 2 pastured eggs, whisked
- 3 tablespoons coconut aminos
- 1 small bell pepper, chopped
- 1 onion, chopped
- 3/4 cup celery, chopped
- 1 1/2 teaspoons garlic powder
- 1 1/2 teaspoons ground black pepper
- 1 1/2 teaspoon kosher salt

Directions:

1. Mix all of the listed ingredients in a large bowl.

2. Spread the mixture evenly on a greased loaf pan.

3. Preheat your oven to 375°F. Bake the loaf for about an hour.

4. Remove from oven and set aside to cool for a few minutes.

5. Serve warm.

Grilled Portobello Mushroom Burgers

Cooking Time: 1 hour 30 minutes
Serves: 6

Ingredients:

- 6 whole wheat buns, toasted
- 6 large Portobello mushrooms caps of about 5 inches diameter, cleaned
- 2 cloves garlic, minced
- 6 slices red onion
- 6 slices tomatoes
- 3 bib lettuce leaves, halved

- 1/2 cup balsamic vinegar
- 3 tablespoons olive oil
- 1 1/2 tablespoons sugar
- 3/4 cup water
- 1/2 teaspoon cayenne pepper

Directions:

1. Place the mushrooms in a glass dish with the stem area facing up.

2. Mix together in a bowl, vinegar, water, sugar, garlic, cayenne pepper and oil. Pour over the mushrooms.

3. Cover and refrigerate for at least an hour. Flip the mushrooms over half way through in the refrigerator.

4. Prepare a charcoal grill or preheat a broiler.

5. Remove the grill rack and spray the rack with cooking spray.

6. Place the rack at least 4 inches away from the heat source.

7. Grill or broil the mushrooms over medium heat. Turn the mushrooms over a couple of times. Grill until the mushrooms are tender. Baste the mushrooms with the remaining marinade.

8. Place a mushroom over each of the buns. Layer tomato and onion slices on top of the mushrooms. Finish with a lettuce leaf and top of the other half of the bun.

Thai-Style Papaya Pork

Cooking Time: 20 minutes
Serves: 4

Ingredients:

- 1/2 pound pork tenderloin, cut into strips
- 1 pound papaya, peeled and cubed
- 3 garlic cloves, minced
- 2 teaspoons low sodium soy sauce
- 1 teaspoon olive oil
- 1/2 teaspoon ginger, grated
- 1/4 cup cilantro, chopped
- 1 lime, juiced
- 1 cup jasmine rice, cooked

Directions:

1. Heat up a pan with the olive oil over medium high heat. Add the pork and cook for 3 minutes.

2. Stir in the soy sauce, ginger and garlic. Cook for 5 minutes.

3. Remove the pan from heat. Add lime juice, cilantro and papaya and toss to coat.

4. Serve with the rice on the side.

Hearty Jambalaya

Cooking Time: 6 hours 15 minutes
Serves: 2

Ingredients:

- 1 can (14 ounce) tomatoes, diced, with its juices
- 1/2 pound fully cooked turkey sausage, cubed
- 1/2 pound chicken breasts, boneless, skinned, cut into 1 inch cubes
- 4 ounce canned tomato sauce
- 1/2 cup onions, diced
- 1/2 a small red bell pepper diced
- 1/2 a small green bell pepper, diced
- 1/2 cup chicken broth
- 1 celery stalk, leaves, chopped
- 1 tablespoon tomato paste
- 1 teaspoon dried oregano
- 1 teaspoon Cajun seasoning
- 1 teaspoon garlic, minced
- 1 bay leaf
- 1/2 pound medium shrimp, cooked
- 1 teaspoon hot sauce
- Hot cooked rice to serve

Directions:

1. Add all the Ingredients into a pressure cooker except the shrimp and rice. Cover.

2. Set the cooker on low for 6-7 hours or until the chicken is cooked through.

3. Add shrimp. Cover and cook again for 15 minutes.

4. Discard bay leaves.

5. Serve hot over the hot rice.

Honey Whole Wheat Bread

Cooking Time: 1 hour 10 minutes
Serves: 4

Ingredients:

- 1 1/2 cups soy flour
- 2 1/2 cups all-purpose flour
- 1 1/2 cups whole wheat flour
- 1/2 cup dry rolled oats
- 6 tablespoons soy flour
- 6 tablespoons flaxseed meal or ground flaxseed
- 1 1/2 tablespoons poppy seeds
- 1 1/2 tablespoons sesame seeds
- 1 teaspoon sea salt
- 2 1/8 tablespoons yeast
- 1/4 cup honey
- 2 tablespoons olive oil
- 1/2 cup unsweetened apple sauce
- 1 1/2 cups water (warm)

Directions:

1. Mix together the warm water and oats,

2. Add the dry Ingredients, except all-purpose flour, to the mixer bowl of the electric mixer. Mix well.

3. Add applesauce, honey and oil and mix with your hands. Add the oat mixture and mix again.

4. Attach the dough attachment to the mixer and run the mixer for about 2-3 minutes.

5. Add flour 1/4 cup at a time and run the mixer until a smooth dough is formed.

6. Place the dough in a bowl and cover. Keep it in a warm place for a couple of hours until it doubles in size.

7. Place the dough onto a lightly floured work area and punch the dough. Divide the dough into 2 and shape into loaves.

8. Place the dough into greased loaf pans. Cover and set aside to rise for a couple of hours.

9. Bake in a preheated oven at 350 degree F for 25 minutes or until golden brown.

10. When done, remove the loaves from the pan and place on a wire rack to cool.

Simple and Delicious Beef Stew

Cooking Time: 1 hour 30 minutes
Serves: 2

Ingredients:

- 3/4 pound beef stew meat
- 4 ounce mushrooms, sliced
- 1 medium sweet potato, peeled, chopped into chunks
- 1 medium onion, chopped
- 1 stalk celery, chopped
- 1 1/2 tablespoons garlic, minced
- 1 tablespoon coconut oil
- 1 tablespoon butter
- 1 bay leaf
- 3 cups beef broth
- 1/2 teaspoon garlic powder
- 1 tablespoon arrowroot powder
- 1/2 tablespoon balsamic vinegar
- Salt to taste
- Pepper powder to taste

Directions:

1. Place a large Dutch oven or saucepan over medium heat. Add coconut oil. When the oil melts, add onions and garlic and sauté until onions are translucent.

2. Sprinkle garlic powder, salt and pepper over meat. Toss to coat well.

3. Meanwhile, place a skillet over medium heat. Add 1/2-tablespoon butter. When butter melts, add meat and cook on both the sides for about a minute each. Remove from the skillet and add it to the Dutch oven.

4. Add 2 cups of beef broth and reduce heat to low.

5. Add sweet potatoes, celery and bay leaf. Stir. Cover and let it simmer.

6. Meanwhile, add remaining butter to the skillet. Add mushrooms and sauté until mushrooms are tender. Add vinegar.

7. Add arrowroot powder to the remaining beef broth. Add this to the pan of mushrooms stirring constantly until thick. Transfer into the Dutch oven. Mix well and simmer for about an hour or until the meat is cooked.

8. Ladle the stew into bowls and serve.

The Green Pasta

Cooking Time: 25 minutes
Serves: 2

Ingredients:

- 6 ounce uncooked, whole wheat pasta
- 3 cups fresh broccoli florets without stem
- 3 cloves garlic, minced
- 2 tablespoons parmesan or Romano cheese, grated
- 1 tablespoon olive oil, divided
- Salt to taste
- Freshly cracked pepper powder to taste

Directions:

1. Place a large pot of water to boil. Add 1-teaspoon salt and pasta.

2. When the pasta is almost cooked, add broccoli. Cook until the pasta is al dente.

3. Drain the pasta. Retain about 1 cup of the cooked water.

4. Return the pot back to heat. Add 1/2 tablespoon oil. When heated, add garlic and sauté until golden.

5. Lower heat and add the drained pasta and broccoli. Mix well.

6. Add the remaining olive oil and cheese. Mix well. Add the retained cooking water, salt and pepper.

7. Heat well and serve hot.

Jamaican Beef Kabobs

Cooking Time: 12 minutes
Serves: 4

Ingredients:

- 12 ounces beef, cubed
- 2 plantains, peeled and chunked
- 1 red onion, cut into wedges
- 2 tablespoons red wine vinegar
- 1 tablespoon vegetable oil
- 1 tablespoon Jamaican seasoning
- Mixed salad greens

Directions:

1. Mix Jamaican seasoning with oil and vinegar in a bowl and stir.

2. Add in the meat, plantain and onion wedges. Toss to coat.

3. Thread the ingredients on skewers and brush them with the remaining marinade.

4. Preheat your grill over medium high heat and place the skewers on it. Cook for beef kabobs for 7 minutes. Then flip them and cook for 7 minutes more.

5. Serve with mixed salad greens as a side dish.

Cheesy Zucchini Pizza

Cooking Time: 15 minutes
Serves: 1

Ingredients:

- 1 whole zucchini
- 1/4 cup part skim mozzarella cheese, shredded
- 1/2 cup pizza sauce
- Ground black pepper to taste
- Salt to taste
- Olive oil cooking spray

Directions:

1. Slice the zucchini into 1/4-inch pieces and spray both sides of each of the slices with olive oil.

2. Sprinkle the slices with salt and pepper.

3. Preheat your broiler to 500°F.

4. Broil the zucchini slices for 2 minutes then flip the sides and broil for another 2 minutes.

5. Transfer the slices to a plate and cover each slice with pizza sauce. Sprinkle each slice with mozzarella cheese.

6. Broil for the slices for a few minutes until the cheese melts.

7. Serve the zucchini pizza hot.

Chicken Tikka Masala

Cooking Time: 4 hours
Serves: 2

Ingredients:

- 1/2 a 15 ounce can crushed tomatoes
- 1 onion, chopped
- 2 cloves garlic
- 1 tablespoon tomato paste
- 1 teaspoon garam masala (Indian spice powder)
- 3/4 pound boneless chicken thighs
- 2 tablespoons cilantro, chopped
- 1/2 tablespoon fresh lemon juice
- 1/2 cup long grain white rice,
- 1/2 teaspoon salt
- 1/4 teaspoon pepper powder
- 1/4 cup low fat cream

Directions:

1. Add crushed tomatoes, onion, garlic, tomato paste, gram masala, salt and pepper to the slow cooker.

2. Lay chicken thighs on top of the vegetables.

3. Set the slow cooker on Low for 7-8 hours or on High for 3-4 hours.

4. Meanwhile, cook rice according to the instructions on the package.

5. When the chicken done add the cream, stir well. Garnish with cilantro and serve with hot rice.

Fresh Mediterranean Quinoa Salad

Cooking Time: 35 minutes
Serves: 2

Ingredients:

- 2 cups quinoa, cooked according to instructions on the package
- 1/2 cup red onions, chopped
- Juice of a lemon
- 1/2 cup kalamata olives, pitted, sliced
- 1/4 cup extra virgin olive oil
- 4 cups cucumber, peeled, diced
- 2 cups cherry tomatoes, quartered
- 1/2 cup feta, crumbled
- Black pepper powder to taste

Directions:

1. Cool the quinoa and transfer into a large bowl. Add onions, olives, cucumber, tomatoes, and lemon juice. Mix well.

2. Pour the olive oil over the salad. Add feta and pepper and toss well. Taste and adjust the seasonings if necessary.

Earl Grey Beef Brisket

Cooking Time: 7 hours 20 minutes
Serves: 6

Ingredients:

Beef Brisket:

- 4 pounds beef brisket
- 8 Earl Gray tea bags
- 1 pound sweet onion, chopped
- 1/2 pound celery, chopped
- 1 pound carrot, chopped
- 4 cups water
- A pinch of salt
- Black pepper to the taste

Sauce:

- 16 ounces canned tomatoes, chopped
- 8 Earl Grey tea bags
- 1 pound sweet onion, chopped
- 1/2 pound celery, chopped
- 4 ounces vegetable oil
- 1 ounce garlic, minced
- 1 cup white vinegar
- 1 cup palm sugar

Directions:

1. Pour the water in a pot. Add 1 pound carrot, 1 pound onion, 1/2 pound celery, salt and pepper. Stir and simmer over medium high heat.

2. Add the beef brisket and 8 Earl Grey tea bags to the pot while stirring. Cover and turn down the heat to medium low. Let it cook for 7 hours.

3. Prepare a pan with vegetable oil and place it over medium high heat. Stir in the remaining chopped onion and sauté for 10 minutes.

4. Add the remaining sauce ingredients including the rest of the tea bags and cook until the vegetables are done. Discard tea bags afterwards.

5. Transfer the beef brisket to a cutting board and set aside to cool down.

6. Slice and serve immediately with the sauce.

Zucchini Pad Thai

Cooking Time: 30 minutes
Serves: 2

Ingredients:

For the sauce:

- 3/4 tablespoon coconut sugar
- 1 teaspoon Sriracha sauce or to taste
- 2 tablespoons tamarind paste
- 2 teaspoons low sodium tamari
- 1 tablespoon lime juice
- 2 tablespoons low sodium chicken stock

For the noodles:

- 1 large carrots, peeled, trimmed with top and bottom sliced off
- 2 large zucchini, trimmed with top and bottom sliced off

For Pad Thai:

- 1 1/2 cups bean sprouts
- 1 large skinless boneless chicken breast, sliced
- 1 egg, beaten
- 2 teaspoons olive oil, divided
- 1 green onion, thinly sliced
- 2 tablespoons peanuts, finely chopped
- Lime wedges to serve
- Salt to taste
- Pepper powder to taste

Directions:

1. To make noodles: Make noodles out of the carrot and zucchini by using a spiralizer or a julienne peeler.

2. For pad Thai: Place a nonstick pan over medium heat. Add 1/2-teaspoon oil. When oil is heated, add egg, salt and pepper. Keep stirring so as to scramble it. Remove from the pan when cooked and place it in a bowl.

3. Place a large nonstick pan over medium heat. Add remaining oil. When oil is heated, add chicken breasts, salt and pepper. Cook until the chicken is tender inside and golden brown outside. Place it along with the egg.

4. To make sauce: Add all the Ingredients of the sauce to a bowl and mix well. Place the pan back on heat. Pour the sauce mixture into the pan and cook until it is bubbly.

5. Add zucchini and carrot noodles and cook sauté for a few minutes until it is thoroughly heated and slightly softened.

6. Add chicken, eggs, and sprouts. Mix well and heat thoroughly.

7. Garnish with lemon wedges, green onion and peanuts and serve immediately.

Fresh Veggie Quesadillas

Cooking Time: 15 minutes
Serves: 2

Ingredients:

- 1 small green pepper, thinly sliced
- 1 small red pepper, thinly sliced
- 1 onion, thinly sliced
- 3 teaspoons olive oil
- 1 teaspoon ground cumin
- Red chili powder to taste
- 3 tablespoons fresh parsley, snipped
- 1/4 cup low fat cream cheese
- 5 flour tortillas
- Salsa

Directions:

1. Place a large nonstick skillet over medium heat. Add a teaspoon oil. When the oil is hot, add onions and bell peppers. Sauté until tender.

2. Add cumin powder and chili powder. Sauté for a minute.

3. Add parsley, mix well, and remove from heat.

4. Spread cream cheese on one half side of each of the tortillas. Sprinkle the pepper mixture over the cream cheese layer.

5. Fold the tortillas over the pepper.

6. Place the folded tortillas on a baking sheet. Brush the tortillas with the remaining oil.

7. Bake in a preheated oven at 425 degrees F for 5-6 minutes.

8. Slice into wedges and serve with salsa.

Whole Wheat Baked Macaroni

Cooking Time: 40 minutes
Serves: 6

Ingredients:

- 1 pound lean ground beef
- 14 ounces whole-wheat elbow macaroni, cooked
- 30 ounces (2 jars) low sodium spaghetti sauce
- 3/4 cup Parmesan cheese
- 1 onion, diced

Directions:

1. Sauté onions on a large pan over medium heat for a couple of minutes until the onions become translucent.

2. Add the ground beef to the pan and cook until brown.

3. Add the macaroni and spaghetti sauce to the pan and mix well.

4. Preheat your oven to 350°F. Place macaroni mixture on a greased baking dish and transfer the dish to the oven. Let it bake for about 30 minutes.

5. Serve to table and garnish with Parmesan cheese.

Honey Garlic Chicken Wings

Cooking Time: 1 hour 20 minutes
Serves: 4

Ingredients:

- 2 pounds chicken wings
- 2 tablespoons honey
- 3 garlic cloves, minced
- 1 and 1/4 cups balsamic vinegar
- 2 tablespoons olive oil
- 1 tablespoon Italian seasoning
- Black pepper to the taste
- A pinch of salt

Directions:

1. Combine the chicken wings with olive oil, salt, pepper and Italian seasoning in a bowl. Toss to coat.

2. Transfer the wings to a baking dish. Bake in a preheated oven at 425°F and bake for an hour.

3. Prepare a pan and heat it up over medium heat. Stir in the honey, garlic, and vinegar stir. Boil and simmer the glaze for 10 minutes.

4. Take out the chicken wings and transfer them to a bowl. Add the glaze and toss to coat. and return them to baking dish.

5. Place the wings back on the baking dish and back to the oven again. Let them bake for 5 more minutes.

6. Serve while hot.

Roasted Vegetable Farro Salad

Cooking Time: 45 minutes
Serves: 4

Ingredients:

- 1 1/2 cups dry farro grain, cook according to package instructions
- 1 onion, peeled, cut into wedges
- 1 1/2 cups broccoli florets
- 1 1/2 cups cauliflower florets
- 1 tablespoon extra-virgin olive oil
- 1/2 cup dried cranberries
- 1/2 cup hazelnuts, chopped
- 2 tablespoons fresh parsley
- Salt to taste
- Pepper powder to taste

For the dressing:

- Zest of a lemon, grated
- 1/4 cup extra virgin olive oil
- 2 tablespoons lemon juice
- 2 tablespoons apple cider vinegar
- 1/4 teaspoon sea salt

Directions:

1. Add all the dressing Ingredients to a bowl and whisk well. Set it aside.

2. Drizzle olive oil over the cauliflower, broccoli, and onion. Sprinkle with salt and pepper.

3. Roast vegetables in a preheated oven at 450 degree F until the vegetables are brown.

4. Remove the vegetables from the oven and let it cool naturally.

5. Cool the cooked farro. Add the cooked vegetables to farro. Mix in the rest of the ingredients.

6. Pour dressing and toss well. Serve chilled or at room temperature.

Healthy Brown Rice Pilaf

Cooking Time: 45 minutes
Serves: 4

Ingredients:

- 2 1/4 cup brown rice, rinsed, drained
- 4 cups water
- 1 1/2 teaspoon salt
- 1/4 teaspoon saffron threads
- 1/4 teaspoon ground turmeric
- 1 teaspoon orange zest
- 6 tablespoons fresh orange juice
- 3 tablespoons canola oil
- 1/2 cup pistachio nuts, chopped
- 1/2 cup dried apricots, chopped

Directions:

1. Place a large saucepan over high heat. Add rice, water, 1/2-teaspoon salt, saffron, and turmeric. Cook until the rice is done.

2. Transfer the cooked rice to a large bowl.

3. Mix together orange zest, orange juice, oil and remaining salt. Whisk well. Pour this over the cooked rice. Add nuts and apricots and toss well.

4. Serve immediately.

Soba Noodles with Mushroom

Cooking Time: 30 minutes
Serves: 4

Ingredients:

- 1/4 cup canola oil
- 2 shallots, minced
- 2 carrots, finely chopped
- 4 cloves garlic, minced
- 3 tablespoon fresh ginger, minced
- 17 1/2 0unces white or brown mushrooms, sliced
- 2 cups frozen edamame, thawed
- 3 cups low sodium broth (vegetable or chicken)
- 1/4 cup low sodium soy sauce
- 2 teaspoons lemon zest, grated
- 2 cups spinach, rinsed, chopped
- 1 cup firm tofu, cut into 1/2 inch pieces
- 1/2 teaspoon freshly ground pepper
- 12 1/3 ounces soba noodles

Directions:

1. Bring a large pot of water to boil. Add the soba noodles. Cook until al dente.

2. Drain. Rinse with cold water. Set aside.

3. Meanwhile place a large nonstick pan or wok over medium heat. Add canola oil.

4. Add shallot, ginger and garlic and sauté for a minute. Add carrot and mushrooms. Stir well.

5. Reduce heat, cover, and cook until the mushrooms are soft.

6. Uncover and increase the heat to medium. Add edamame and sauté until thoroughly heated.

7. Add broth, soy sauce, and lemon zest and bring to a boil.

8. Add spinach. Sauté until spinach wilts. Add tofu, salt, and pepper. Stir and remove from heat.

9. Add the boiled soba noodles and toss well

10. Serve hot in bowls.

DASH Chicken Noodles Soup

Cooking Time: 35 minutes
Serves: 8

Ingredients:

- 2 cups chicken meat, cooked
- 1 pound egg noodles
- 6 cups low sodium chicken stock
- 4 cups green onions, chopped
- 3 celery stalks, thinly sliced
- 3 carrots, sliced
- 1 onion, chopped
- 1/3 cup water
- 1 tablespoon olive oil
- 1 tablespoon cornstarch
- 2 teaspoons parsley, chopped
- 2 teaspoons thyme, chopped
- A pinch of salt
- Black pepper to the taste

Directions:

1. Prepare a pot with olive oil and place it over medium high heat. Stir in the chopped onions, celery and carrots and cook for about 6 minutes.

2. Stir in the chicken stock and boil. Once boiling, reduce the heat to medium low and simmer for 20 minutes.

3. Prepare another bowl and stir the water with cornstarch.

4. Pour the cornstarch mixture over the soup and add the noodles. Stir and cook for a couple of minutes.

5. Stir in the chicken meat, parsley, thyme, salt and pepper and let it cook for 3-4 minutes.

6. Serve and sprinkle the soup with sliced green onions.

Spinach Artichoke Angel Hair Pasta with Shrimp

Cooking Time: 25 minutes
Serves: 2

Ingredients:

- 6 ounces angel hair pasta, broken in half
- 1/2 cup canned artichokes, quartered
- 1/2 cup raw shrimps, peeled and deveined
- 1/2 cup mushrooms, thinly sliced
- 2 1/2 cups chicken broth
- 1 onion, diced small
- 2 garlic cloves, thinly sliced
- 1 tablespoon grape seed oil
- 1/2 cup fresh spinach, roughly chopped
- 1/2 teaspoon dried oregano
- Salt to taste
- Pepper to taste
- Pinch of crushed red pepper flakes (optional)

Directions:

1. Pour grade seed oil on a pot over medium heat. Wait for the oil to heat up.

2. Add mushrooms, onions, and garlic to the pot and sauté them for a few minutes.

3. Pour the chicken broth into the pot. Add in the pasta, artichokes, shrimps, oregano, salt, and pepper and optionally, the red pepper flakes.

4. Mix all of the ingredients and bring to a boil. Cover and keep cooking until the pasta slightly undercooked (al dente).

5. Add the chopped spinach and continue for a couple of minutes until the spinach wilts.

6. Add any remaining seasonings if necessary and serve hot.

Grilled Chicken and Ratatouille

Cooking Time: 1 hour 10 minutes
Serves: 2

Ingredients:

- 1 1/2 pounds boneless chicken breast halves (about 3 pieces)
- 1 medium eggplant, halved lengthwise
- 2 medium red onions, cut into 1 inch wide wedges (Do not remove the root part)
- 1 medium red bell pepper, cut into 1 inch wide strips
- 1 medium zucchini, halved lengthwise
- 1 medium tomato, halved crosswise
- 1/4 cup fresh basil, thinly sliced
- 1 tablespoon olive oil
- 1 teaspoon red wine vinegar

Directions:

1. Mix together zucchini, eggplant, bell pepper, onion and tomatoes in a bowl. Season with salt and pepper. Pour olive oil over it and toss well.

2. Preheat a broiler or prepare a high heat barbecue. Place the vegetables on the grill.

3. Grill the vegetables until it is slightly charred. Different vegetables char at different times, remove as it becomes ready.

4. Placed the grilled vegetables on your cutting board.

5. Add chicken into the same bowl. Turn the chicken around to coat the chicken with the remaining oil. Season with salt and pepper.

6. Place the chicken on the grill, cover and cook on both the sides until tender. Let it stand for a while.

7. When the vegetables are cool enough to handle, coarsely chop the vegetables and place in a bowl. Add basil and vinegar and mix well.

8. Place the chicken on your cutting board. Slice the chicken crosswise into 1/2 inch thick slices.

9. Serve chicken with the grilled vegetables.

Summer Vegetable Rice

Cooking Time: 25 minutes
Serves: 2

Ingredients:

- 1 tablespoon vegetable oil
- 2 onions chopped
- 2 teaspoons ground ginger
- 1 teaspoon turmeric powder
- 1 teaspoon cumin seeds
- 4 cups mixed vegetables, chopped (vegetables of your choice like cauliflower, carrots, peas, etc.)

- 2 large potatoes, peeled, diced
- 2 cups brown rice, soaked in water for at least an hour
- 1 teaspoon salt
- 5 cups water
- 1 can kidney beans

Directions:

1. Heat a large skillet over medium heat.

2. Add oil. When oil is hot, add cumin and heat until fragrant. Add onion. Sauté for a couple of minutes. Add ginger and turmeric, sauté for a minute.

3. Add rest of the Ingredients stir and bring to a boil.

4. Lower heat. Cover, and cook until rice is done.

5. Serve hot.

Vegan Lentil Soup

Cooking Time: 30 minutes
Serves: 4

Ingredients:

- 1 cup dry lentils
- 5 cups water
- 2 carrots, peeled and chopped
- 2 celery stalks, diced
- 1 onion, chopped
- 1 can unsalted tomatoes, drained and diced

- 1 tablespoon olive oil
- 1 teaspoon dried thyme
- 1/2 teaspoon garlic powder
- Ground black pepper
- A pinch of salt

Directions:

1. Pour the olive oil in a large Dutch oven or pot and place it over medium heat. Wait for the oil to simmer and then add the carrots, chopped onions and celery. Continue cooking and stirring until the onion softens and starts to turn translucent.

2. Stir in the garlic powder, thyme, and black pepper. Let it cook and keep stirring until fragrant.

3. Add in the tomatoes and cook for a couple more minutes while stirring.

4. Add the water, lentils, and salt. Increase the heat and wait until boiling. Then partially cover the pot and decrease the heat to let it simmer gently. Cook for about 30 minutes or until lentils are tender while their shapes are still intact.

5. Serve immediately.

Shiitake Mushroom Chili

Cooking Time: 45 minutes
Serves: 2

Ingredients:

- 4 ounces shiitake mushrooms, sliced
- 3/4 pound white button mushrooms, sliced
- 9 1/2 ounces canned white kidney beans, rinsed and drained
- 7 1/4 ounces canned stewed tomatoes
- 1/2 cup onions, chopped
- 1/2 tablespoon garlic, minced
- 1/4 cup sliced ripe olives
- 1 tablespoon vegetable oil
- 1/2 cup water
- 2 tablespoons cheddar cheese, shredded
- 1 tablespoon chili powder
- 1/2 teaspoon ground cumin

Directions:

1. Pour vegetable oil on a saucepan over medium heat. Wait for the oil to heat up.

2. Add onions and garlic and sauté until the onions become translucent.

3. Add the ground cumin and chili powder keep sautéing for a few seconds.

4. Add the mushrooms and continue sautéing until the mushrooms become tender.

5. Add water, tomatoes, olives, and white beans. Simmer for about 10 minutes.

6. Garnish with shredded cheddar cheese and serve hot.

Basil Pesto Stuffed Mushrooms

Cooking Time: 30 minutes
Serves: 4

Ingredients:

- 20 cremini mushrooms, washed and stems removed
- Topping:
- 1 1/2 cups panko breadcrumbs
- 1/4 cup melted butter
- 3 tablespoons chopped fresh parsley

Filling:

- 2 cups fresh basil leaves
- 1/4 cup fresh Parmesan cheese
- 2 tablespoons pumpkin seeds
- 1 tablespoon olive oil
- 1 tablespoon fresh garlic
- 2 teaspoons lemon juice
- 1/2 teaspoon kosher salt

Directions:

1. Preheat the oven to 350 F.

2. Line the mushroom caps, rounded side down, onto a baking sheet.

3. In a small bowl, combine the panko, butter and parsley; set aside and prepare the stuffing.

4. Place the basil, cheese, pumpkin seeds, oil, garlic, lemon juice and salt in a food processor. Pulse until the ingredients are evenly mixed, but not pureed.

5. Stuff the mushroom caps with the basil pesto filling.

6. Sprinkle each mushroom with about 1 teaspoon of panko topping.

7. Pat down the topping and bake for 10 to 15 minutes or until golden brown.

Chipotle Spiced Shrimp

Cooking Time: 20 minutes
Serves: 4

Ingredients:

- 3/4 pound uncooked shrimp, peeled and deveined (about 48 shrimp)
- 2 tablespoons tomato paste
- 1 1/2 teaspoons water
- 1/2 teaspoon extra-virgin olive oil
- 1/2 teaspoon minced garlic
- 1/2 teaspoon chipotle chili powder
- 1/2 teaspoon fresh oregano, chopped

Directions:

1. In cold water, rinse shrimp.

2. Pat dry with a paper towel. Set aside on a plate.

3. Whisk together the tomato paste, water and oil in a small bowl to make the marinade. Add garlic, chili powder and oregano and mix well.

4. Spread the marinade (it will be thick) on both sides of the shrimp using a brush and place in the refrigerator.

5. Heat a gas grill or broiler, or prepare a hot fire in a charcoal grill.

6. Coat the grill rack or broiler pan with cooking spray lightly.

7. Put the cooking rack 4 to 6 inches from the heat source.

8. Thread the shrimp onto skewers or lay them in a grill basket, to place on the grill.

9. After 3 to 4 minutes turn the shrimp.

10. When the shrimp is fully cooked, take it off the heat and serve immediately.

DASH Beef Stew

Cooking Time: 7 hours
Serves: 6

Ingredients:

- 2 pounds beef stew, cubed
- 4 cups low sodium beef stock
- 3 garlic cloves, minced
- 2 cups baby carrots
- 2 celery ribs, chopped
- 1 yellow onion, chopped
- 1 cup peas
- 1 cup corn
- 2 tablespoons reduced sodium Worcestershire sauce
- 6 ounces tomato paste
- 1 tablespoon parsley, dried
- 1 teaspoon oregano, dried
- 1/4 cup white flour
- 1/4 cup water
- Black pepper to the taste

Directions:

1. Add all of the ingredients except the flour and water into your slow cooker. Stir well and cook it on "High" setting for 6 hours and 30 minutes.

2. Combine the flour and water in a separate bowl and stir well.

3. Pour the mixture into the stew and let it slow cook on the same setting for 30 minutes more.

4. Serve immediately.

Easy Coconut Shrimp

Cooking Time: 25 minutes
Serves: 2

Ingredients:

- 1/4 cup sweetened coconut
- 1/4 cup panko breadcrumbs
- 1/2 teaspoon kosher salt
- 1/2 cup coconut milk
- 12 large shrimp, peeled and deveined

Directions:

1. Heat the oven to 375 F.

2. Coat a baking sheet lightly with cooking spray.

3. Put the coconut, panko and salt in a food processor and pulse.

4. In a small bowl pour in the coconut milk.

5. Dip each shrimp first into the coconut milk and then into the panko mixture. Place on to the greased baking sheet.

6. Lightly spray the tops of the shrimp with cooking spray.

7. Bake about 10 to 15 minutes until golden brown.

Grilled Eggplant with Toasted Spices

Cooking Time: 25 minutes

Serves: 2

Ingredients:

- 1 large eggplant (aubergine), about 1 1/2 pounds
- 1 teaspoon mustard seed
- 1/2 teaspoon ground cumin
- 1/2 teaspoon ground coriander
- 1/2 teaspoon curry powder
- Pinch of ground ginger
- Pinch of ground nutmeg
- Pinch of ground cloves
- 1 tablespoon olive oil
- 1/2 yellow onion, finely chopped
- 2 cups cherry tomatoes, halved, or 1 cup tomato sauce
- 1 tablespoon light molasses
- 1 garlic clove, minced
- 1 teaspoon red wine vinegar
- 1/4 teaspoon salt
- 1/4 teaspoon freshly ground black pepper
- 1 tablespoon chopped fresh cilantro (fresh coriander)

Directions:

1. Use a charcoal grill or preheat a gas grill or broiler. Put the cooking rack 4 to 6 inches from the heat source.

2. Coat the grill rack or broiler pan with cooking spray.

3. Trim the eggplant and cut lengthwise into 1/4-inch thick slices.

4. Place the slices onto the rack and grill or broil for about 5 minutes on each side, turning once halfway through, until the eggplant is tender and browned. Leave the eggplant into the oven to keep warm.

5. Combine the first 7 spices in in a small bowl.

6. Pour the olive oil into a large frying pan and heat over medium heat until hot but not smoking. Add the spice mixture and cook until fragrant, about 30 seconds.

7. Add the onions and sauce and sauté until onions are soft and translucent.

8. Add the tomatoes, molasses, garlic and vinegar.

9. Cook the sauce, stirring occasionally, until thickened, about 4 minutes.

10. Season with the salt and pepper.

11. Place the eggplant on to a serving dish, pour the sauce over, and garnish with the cilantro.

Fresh Tomato Crostini

Cooking Time: 45 minutes
Serves: 2

Ingredients:

- 4 plum tomatoes, chopped
- 1/4 cup minced fresh basil
- 2 teaspoons olive oil
- 1 clove garlic, minced
- Freshly ground pepper
- 1/4 pound crusty Italian peasant bread, cut into 4 slices and toasted

Directions:

1. In a medium bowl mix together the tomatoes, basil, oil, garlic and pepper.

2. Cover and marinate for 30 minutes.

3. Spoon the tomato mixture onto crispy toast.

4. Serve at room temperature.

Pork and Shiso Soup

Cooking Time: 20 minutes
Serves: 3

Ingredients:

- 1/2 pounds pork, thinly sliced
- 8 Shiso leaves, stems removed
- 4 tomatoes, chopped
- 1 eggplant, chopped
- 1 yellow onion, chopped
- 3 teaspoons low sodium fish sauce
- 4 cups water

Directions:

1. Set up a pot over medium high heat and add the onions. Stir and cook for about 3 minutes.

2. Add the pork slices. Stir and cook until they all brown.

3. Pour in the water and keep stirring and cooking until the soup boils.

4. Stir in chopped tomatoes and eggplants and keep it boiling for a few more minutes.

5. Reduce heat to medium and let it simmer for 15 minutes.

6. Add the remaining ingredients while still stirring. Serve into separate bowls.

Ginger-Marinated Grilled Portobello Mushrooms

Cooking Time: 15 minutes
Serves: 2

Ingredients:

- 4 large Portobello mushrooms
- 1/4 cup balsamic vinegar
- 1/2 cup pineapple juice
- 2 tablespoons chopped fresh ginger, peeled
- 1 tablespoon chopped fresh basil

Directions:

1. Clean mushrooms with a damp cloth and remove the stems.

2. Whisk together the vinegar, pineapple juice and ginger in a small bowl. Spread the marinade over the mushrooms.

3. Cover and refrigerate the mushrooms for an hour, flipping the mushrooms over once half way through.

4. Use a charcoal grill, or heat a gas grill or broiler. Lightly coat the grill rack or broiler pan with cooking spray. Fit the cooking rack 4 to 6 inches from the heat source.

5. Grill or broil the mushrooms, turning often, until tender, about 5 minutes on each side. Baste with marinade to keep the mushrooms from drying out.

6. Transfer the mushrooms by tongs to a serving platter. Garnish with basil before serving.

Marinated Portobello Mushrooms with Provolone

Cooking Time: 25 minutes
Serves: 2

Ingredients:

- 2 Portobello mushrooms, stemmed and wiped clean
- 1/2 cup balsamic vinegar
- 1 tablespoon brown sugar
- 1/4 teaspoon dried rosemary
- 1 teaspoon minced garlic
- 1/4 cup grated (1 ounce) provolone cheese

Directions:

1. Preheat the broiler or grill. Set up the rack 4 inches from the heat source.

2. Coat glass baking dish with cooking spray. Put the mushrooms in the dish, gill side up.

3. Whisk together the vinegar, brown sugar, rosemary and garlic in a small bowl.

4. Pour the mixture over the mushrooms and wait for 5 to 10 minutes to marinate.

5. Broil or grill the mushrooms, turning once, until they're tender, for about 4 minutes on each side.

6. Sprinkle grated cheese over each mushroom and continue to broil or grill until the cheese melts.

7. Serve immediately.

Whole Wheat Pizza Margherita

Cooking Time: 1 hour 25 minutes
Serves: 2

Ingredients:

Whole-wheat pizza dough:

- 3/4 cup whole-wheat flour
- 2 tablespoons barley flour
- 1 teaspoon active dry yeast
- 1 tablespoon oats
- 2 teaspoons gluten
- 3/4 cup warm water
- 1 tablespoon olive oil

Toppings:

- 2 ounces fresh mozzarella
- 2 1/2 cups tomatoes, sliced
- 2 1/2 cups spinach, chopped
- 1/4 cup basil, chopped
- 1 tablespoon oregano, minced
- 1 tablespoon garlic, minced
- 1 teaspoon black pepper

Directions:

1. Activate the yeast according the product's instructions.

2. Combine all dry ingredients together.

3. Add the yeast mixture and olive oil. Knead dough for 10-15 minutes until you get the ideal texture.

4. Place the dough in refrigerator and leave it to rise for at least 1 hour.

5. Preheat your oven to 450°F.

6. Roll out the pizza dough on to a floured work surface ensuring it is about 1/4-inch thick.

7. Place dough on a pizza peel or baking sheet.

8. Add all of the toppings ingredients.

9. Bake pizza for approximately 10-12 minutes or until the crust is crisp and cheese melted.

Mustard-coated Pork Tenderloin

Cooking Time: 15 minutes
Serves: 4

Ingredients:

- 1 pound pork tenderloin
- 3 tablespoons Dijon mustard
- 3 tablespoons honey
- 1 tablespoon olive oil
- 1 teaspoon dried rosemary
- Ground black pepper
- Salt

Directions:

1. Place the rack set in the middle of your oven and preheat it to 425°F.

2. Prepare a bowl and combine the honey, mustard and rosemary in it. Mix really well and then set it aside.

3. Place an oven-safe skillet over high heat and let it preheat for a few minutes.

4. Pour the olive oil and wait for the skillet to heat up further. While heating up, pat the pork tenderloin dry with a paper towel and lightly season all sides of the meat with salt and pepper.

5. Once the skillet is hot, spend the next 12 minutes searing each side of the pork tenderloin until golden brown.

6. Remove from the skillet from the heat source and spread the honey-mustard mixture evenly until the pork tenderloin is fully coated.

7. Put the skillet back in the oven and cook for more 15 minutes or stop cooking until the temperature hits 145°F.

8. Remove the pork tenderloin from the oven. Slice and serve after 3 minutes.

Roasted Butternut Squash

Cooking Time: 30 minutes
Serves: 2

Ingredients:

- 1 medium butternut squash
- 1 tablespoon olive oil
- 1 tablespoon chopped fresh thyme
- 1 tablespoon chopped fresh rosemary
- 1/2 teaspoon salt

Directions:

1. Preheat oven to 425 F.

2. Coat a baking sheet lightly with non-stick cooking spray.

3. Peel skin from butternut squash and cut into chunks, about 1/2 inch wide and 3 inches long.

4. Add the squash, oil, thyme, rosemary, and salt in a medium bowl and mix the ingredients until the squash is evenly coated.

5. Spread the mixture onto the baking sheet and roast for 10 minutes.

6. Remove the baking sheet from the oven.

7. Shake to loosen the squash.

8. Put it in the oven again and continue to roast for another 5 to 10 minutes until golden brown.

Shrimp Marinated In Lime Juice and Dijon Mustard

Cooking Time: 1 hour 15 minutes
Serves: 2

Ingredients:

- 1 medium red onion, chopped
- 1/2 cup fresh lime juice, plus lime zest as garnish
- 2 tablespoons capers
- 2 tablespoons Dijon mustard
- 1/2 teaspoon hot sauce
- 1 cup water
- 1/2 cup rice vinegar
- 3 whole cloves
- 1 bay leaf
- 1 pound uncooked shrimp, peeled and deveined

Directions:

1. Mix the onion, lime juice, capers, mustard and hot sauce in a shallow baking dish and set aside.

2. Add water, vinegar, cloves and bay leaf into a large saucepan.

3. Bring to a boil and add the shrimp.

4. Stir constantly and cook for 1 minute.

5. Drain and transfer the shrimp to the shallow dish containing the onion mixture.

6. Make sure to discard the cloves and bay leaf. Stir to combine.

7. Cover and keep in a refrigerator until well chilled for about 1 hour.

8. Divide the shrimp mixture among individual small bowls and garnish each with lime zest.

9. Serve cold.

Baked Chicken with Spanish Rice

Cooking Time: 15 minutes
Serves: 2

Ingredients:

- 3 1/4 cup skinless and boneless chicken breast, cooked and diced
- 5 cups white rice, cooked in unsalted water
- 2 teaspoons vegetable oil
- 1 cup tomato sauce
- 3/4 cup sweet green peppers
- 1 cup onions, chopped
- 1 teaspoon parsley, chopped
- 1 1/2 teaspoon garlic, minced
- 1/4 teaspoon black pepper

Directions:

1. Sauté green peppers and onions in a large skillet for about 5 minutes on medium heat.

2. Add the spices and tomato sauce and mix while bringing to a boil.

3. Add chicken breasts and continuing heating.

4. Serve over cooked rice.

Pork and Peach Salad with Walnuts

Cooking Time: 10 minutes
Serves: 4

Ingredients:

- 1 pound pork tenderloin, cubed
- 1 peach, pitted and sliced
- 1/4 cup walnuts
- 10 ounce packaged fresh spinach leaves
- 1 tablespoon olive oil
- Balsamic vinegar

Directions:

1. Prepare a nonstick skillet and pour olive oil on it. Set the skillet over medium-high heat.

2. Add in the pork and cook one side for 3 minutes or until the side is browned. Repeat this step for the other sides. Set the skillet aside afterwards.

3. Prepare a serving plate and cover it with spinach leaves. Place the peach slices on top and then top the dish off with the pork tenderloin.

4. Sprinkle the salad with and pour in some balsamic vinegar. Serve immediately.

Delicious Coconut Quinoa Curry

Cooking Time: 3 hours 15 minutes
Serves: 2

Ingredients:

- 1 medium sweet potato, peeled + chopped (about 3 cups)
- 2 cups of fresh green beans (cut into 1/2 inch pieces)
- 1 medium size carrot (cut into small bit size pieces)
- 1/2 white onion, diced (about 1 cup)
- 1 (15 oz) can organic chickpeas, drained and rinsed
- 1 (28 oz) can diced tomatoes
- 2 (14.5 oz) cans coconut milk (either full fat or lite)

- 1/4 cup quinoa
- 2 garlic cloves, minced (about 1 tablespoon)
- 1 tablespoon freshly grated ginger
- 1 tablespoon freshly grated turmeric (or 1 teaspoon ground)
- 2 teaspoon tamari sauce
- 1/2 - 1 teaspoon chili flakes
- 1 - 11/2 cups water

Directions:

1. Pour 1 cup of water into a slow cooker. Put all ingredients to the slow cooker.

2. Stir to incorporate everything fully.

3. Turn slow cooker to high and cook for 3 - 4 hours until sweet potato cooks through and the curry has thickened.

4. Serve as a vegetarian soup or over rice.

Roasted Brussels Sprouts and Chicken Dinner

Cooking Time: 30 minutes
Serves: 2

Ingredients:

- 1 pound boneless skinless chicken breasts cut into 4 pieces
- 4 cups trimmed and quartered Brussels sprouts
- 3 cups Yukon gold or red potatoes, cut into bite-size pieces
- 1 cup medium-diced onions
- 1/3 cup vinaigrette dressing (not fat-free)
- Juice of 1 medium lemon
- 2 teaspoons Dijon mustard
- 1 1/2 teaspoons oregano
- 1/4 teaspoon garlic salt
- 1/4 cup quartered Kalamata olives
- Freshly ground black pepper

Directions:

1. Preheat oven to 400 F.

2. Place the chicken in a single layer in the middle of a sheet pan.

3. Place the Brussels sprouts in one section of the pan, the potatoes in another, and the onions in the remaining space.

4. In a small bowl combine the vinaigrette, lemon juice, mustard, oregano and garlic salt. Drizzle over the chicken and vegetables.

5. Sprinkle with olives and pepper and toss vegetables to coat evenly.

6. Bake for 20 minutes until chicken is cooked through. Place chicken to a plate and stir the vegetables.

7. Roast the vegetables for 15 minutes, until the outer leaves of the Brussels sprouts are crispy and the potatoes are fork tender.

Simple and Delicious Chicken Salad

Cooking Time: 10 minutes
Serves: 2

Ingredients:

- 3 1/4 cups chicken, cooked, cubed, skinless
- 1/4 cups celery, chopped
- 1 tablespoon lemon juice
- 1/2 teaspoon onion powder
- 3 tablespoon light mayo or miracle whip

Directions:

1. Bake cubed chicken and refrigerate.

2. Combine all ingredients with chilled chicken and mix well

Cashew Chicken Salad

Cooking Time: 30 minutes
Serves: 4

Ingredients:

- 1 whole chicken, chopped
- 1 cup cashews, toasted and chopped
- 8 black tea bags
- 4 scallions, chopped
- 2 celery ribs, chopped
- 1 cup mandarin orange, chopped
- 1/2 cup fat free yogurt
- Ground black pepper to the taste
- Water

Directions:

1. Prepare a pot with water and tea bags and add in the chicken pieces. Add more water if necessary until the pieces are fully submerged.

2. Bring the pot to a boil over medium heat. Let it cook for 25 minutes or until the chicken is tender. Once tender, place the chicken on a cutting board to cool down.

3. Transfer 4 ounces of liquid to a bowl. Drain the remaining liquid and discard the tea bags.

4. Once chicken is cool, debone it and shred the meat. Place the shredded meat in the bowl where you poured the liquid earlier.

5. Add in the orange pieces, celery, cashews, and scallions and toss the ingredients.

6. Add the rest of the ingredients and toss until well coated. Refrigerate and serve chilled afterwards.

Simple and Delicious Shepherd's Pie

Cooking Time: 45 minutes
Serves: 4

Ingredients:

- 2 large baking potatoes, peeled and diced
- 1/2 cup low-fat milk
- 1 pound lean ground beef
- 1 medium onion, chopped
- 1 clove garlic, minced
- 2 tablespoons flour
- 4 cups frozen mixed vegetables
- 3/4 cup reduced sodium beef broth
- 1/2 cup shredded cheddar cheese
- Ground pepper to taste

Directions:

1. Put diced potatoes in saucepan and add enough water to cover. Bring to a boil.

2. Turn down the heat and simmer the potatoes, covered, until soft (about 15 minutes). Drain potatoes and mash. Mix in milk, and set mixture aside.

3. Preheat oven to 375 degrees.

4. In a large skillet brown meat, onion, and garlic.

5. Stir in flour, and cook for 1 minute, stirring constantly. Pour in broth. Cook until thick and bubbly, stirring occasionally.

6. Spoon the mixture into an 8 inch square baking dish. Spread the mashed potatoes over vegetable/meat mixture. Sprinkle cheese on top.

7. Bake 25 minutes, until hot and bubbly.

Beef and Bean Jalapeño Chili

Cooking Time: 25 minutes
Serves: 4

Ingredients:

- 1 pound lean ground beef
- 4 cups canned kidney beans, rinsed and drained
- 2 cups canned tomatoes
- Jalapeños, seeded and chopped
- 1 1/2 tablespoons chili powder
- 2 tablespoons cornmeal
- 1 teaspoon coconut sugar
- 1 cup chopped celery
- 1/2 cup chopped onion
- Water (as needed)

Directions:

1. Cook the onions and ground beef in a medium sized pot until the onions become translucent and the meat browns up. Drain grease afterwards.

2. Add in the celery, kidney beans, tomatoes, chili powder and sugar.

3. Cover the pot and continue cooking for about 10 minutes.

4. Pour in as much water to the mixture as you like until you get the preferred consistency.

5. Stir in the cornmeal.

6. Simmer on the stove until all the flavors melded.

7. Pour into a bowl and garnish with chopped jalapeños. Serve immediately.

Thousand Island Tuna Melt

Cooking Time: 15 minutes
Serves: 2

Ingredients:

- 2 whole-wheat English muffins, split
- 6 ounces canned white tuna in water, drained
- 1/4 cup low fat Thousand Island fat free dressing
- 3 ounces low-fat cheddar cheese, grated
- 1/4 cup onion, chopped
- 1/3 cup celery, chopped
- Salt to taste
- Black pepper to taste

Directions:

1. Preheat your broiler.

2. Mix the salad dressing, tuna, onion and celery together.

3. Season the mixture with salt and pepper.

4. Toast the English muffins.

5. Lay down the English muffins on a baking sheet with the split side facing up.

6. Add 1/4 of your tuna mixture to the split side.

7. Broil for about 2-3 minutes.

8. Add the cheese and place the muffins back to the broiler until the cheese is bubbly and melted.

Baked Almond Chicken

Cooking Time: 30 minutes
Serves: 4

Ingredients:

- 4 chicken breasts, skinless and boneless
- 1/2 cup almond meal
- 1 teaspoon ground paprika
- 1/2 teaspoon black pepper
- 1/8 teaspoon salt

Directions:

1. Preheat the oven to 350°F.

2. Place a single chicken breast in a resealable bag filled with all the ingredients. Seal the bag and shake it until the ingredients fully coat the chicken breast.

3. Repeat the previous step with the other chicken breasts. Once done, place the coated chicken breasts on a glass baking dish.

4. Place the dish in the preheated oven and let the chicken bake for about 25 to 30 minutes. Observe the chicken while baking making sure there are no pink spots in the middle and you see clear juices. If you have an instant-read thermometer, stop baking as soon as the center hits 165°F.

5. Serve hot.

Spicy Roasted Broccoli

Cooking Time: 30 minutes
Serves: 8

Ingredients:

- 1 1/4 pounds broccoli, large stems trimmed and cut into 2-inch pieces (about 8 cups)
- 4 tablespoons olive oil, divided
- 1/2 teaspoon salt-free seasoning blend
- 1/4 teaspoon freshly ground black pepper
- 4 cloves garlic, peeled and minced
- 1/4 teaspoon crushed red pepper flakes

Directions:

1. Preheat the oven to 450 F.

2. In a large bowl, toss together the broccoli and 2 tablespoons olive oil. Sprinkle with salt-free seasoning and pepper.

3. Transfer to a rimmed baking sheet and bake for 15 minutes. Mix together 2 tablespoons olive oil, the garlic, and the red pepper flakes.

4. When the broccoli has cooked for about 15 minutes, drizzle the garlic oil over the broccoli and shake the baking sheet to coat the broccoli. Return to the oven and continue baking until the broccoli starts to crisp, about 8 more minutes. Serve while hot.

Mexican Bake

Cooking Time: 50 minutes
Serves: 6

Instructions:

- 1 1/2 cups cooked brown rice
- 1 pound skinless, boneless chicken breast, cut in bite-sized pieces
- 2 (14.5 ounce) cans no-salt-added crushed tomatoes
- 1 (15 ounce) can no-salt-added black beans, drained and rinsed
- 1 cup frozen yellow corn kernels
- 1 cup chopped red bell pepper
- 1 cup chopped poblano pepper
- 1 tablespoon chili powder
- 1 tablespoon cumin
- 4 garlic cloves, crushed
- 1 cup shredded reduced-fat Monterey Jack cheese
- 1/4 cup jalapeno pepper slices (optional)

Directions:

1. Preheat oven to 400 F.

2. Spread rice in a shallow casserole dish. Add chicken on top.

3. In a bowl, combine tomatoes, beans, corn, peppers, seasonings and garlic; pour over chicken. Top with cheese and optional jalapeno.

4. Bake for about 45 minutes or until chicken is fully cooked.

Middle Eastern Hummus Wrap

Cooking Time: 15 minutes
Serves: 1

Ingredients:

- 1/3 cup hummus
- 1 whole-wheat tortillas (8-inch)
- 1 Plum Tomato, thinly sliced
- 1/4 whole Long English Cucumber, thinly sliced
- 1/4 medium red onion, thinly sliced
- 1 cup Baby Lettuce Mix
- 8 leaf fresh mint leaves
- 1 teaspoon 100% Lemon Juice
- 1/2 teaspoon lemon zest
- 1/2 teaspoon ground black pepper

Directions:

1. Spread the hummus onto the tortilla. Top with the remaining ingredients.

2. Roll the tortilla up tightly.

Chicken Veggie Wrap

goPreparation Time: 10 minutes
Serves: 4

Ingredients:

- 4 whole-wheat tortillas
- 6 ounces chicken breasts, cooked and thinly cut into strips
- 2 tomatoes, chopped
- 3/4 cup cucumber, thinly sliced
- 1/3 cup fat-free yogurt
- Black pepper to the taste

Directions:

1. Set up a pan over medium heat. Place a tortilla on top of the pan one by one.

2. Once heated up, set them up flat on a wide surface. Heat up a pan over medium heat, add one tortilla at the time, heat up and arrange them on a working surface.

3. Spread the yogurt on each tortilla.

4. Add the rest of the ingredients to each tortilla and individually roll them up. Serve on separate plates.

Rotisserie Chicken Salad with Creaming Tarragon Dressing

Cooking Time: 20 minutes
Serves: 4

Ingredients:

15 ounces canned white beans
1/3 cup White Balsamic Vinegar
1 tablespoon extra-virgin olive oil
2 garlic clove
2 tablespoon tarragon, fresh, divided
6 cups Salad Mix
1/2 medium red onion, thinly sliced

1 1/3 cups Chicken Breast, Cooked, chopped
12 Grapes (Red or Green), thinly sliced
1 cup English Cucumber, thinly sliced
3 tablespoon pine nuts
3/4 teaspoon ground black pepper

Directions:

1. Add 1/2 cup of the beans, the vinegar, oil, garlic, and 1 tablespoon. of the tarragon to a blender. Cover and puree.

2. Arrange greens on a large platter. Top with remaining beans, onion, chicken, grapes, cucumber, nuts, pepper, and remaining 1 tablespoon tarragon.

3. Serve the dressing on the side.

Spinach-Stuffed Turkey Burger Patties

Cooking Time: 30 minutes
Serves: 4

Ingredients:

- 12 oz Ground Turkey 93% Lean
- 3 cups Fresh Baby Spinach
- 1/3 cup Old Fashioned Rolled Oats
- 1 garlic clove, minced
- 1 egg(s), beaten
- 2 tsp 100% Lemon Juice
- 3/4 tsp ground black pepper
- 1/2 tsp sea salt
- 1 pinch Nutmeg (Ground)

Directions:

1. In a medium bowl, combine all of the ingredients.

2. Spray a large nonstick skillet with cooking spray and place over medium heat. Form mixture into 4 (5-inch wide) patties and cook in the skillet until well done and browned, about 7 to 8 minutes per side.

Ricotta and Pomegranate Bruschetta

Cooking Time: 25 minutes
Serves: 4

Ingredients:

- 6 slice Whole Grain Nut Bread, halved
- 1 cup Low Fat Ricotta Cheese
- 1/2 tsp Grated lemon zest
- 1/2 cup Pomegranate seeds
- 2 tsp Thyme, Fresh
- 1/4 tsp sea salt

Directions:

1. Preheat the oven to 425 F. Place the bread slices onto a large baking sheet. Bake until lightly toasted, about 8 minutes.

2. In a small bowl, stir together the ricotta and lemon zest. Top each piece of toast with the ricotta mixture.

3. Sprinkle with the pomegranate seeds and thyme

Vegetarian Baked Sweet Potatoes

Cooking Time: 20 minutes
Serves: 4

Ingredients:

- 4 sweet potatoes, poked
- 15 ounce canned chickpeas, drained and rinsed
- 1 red bell pepper, cored and diced
- 1/2 red onion, diced
- 1/2 cup nonfat or low-fat plain Greek yogurt
- 1 teaspoon ground cumin
- 1 teaspoon olive oil
- Ground black pepper

Directions:

1. Preheat your oven to 400°F. Place all the poked sweet potatoes in the oven and leave them to bake for about 45 minutes.

2. Pour the yogurt into a small bowl and stir while adding in the black pepper.

3. Prepare a medium-sized pot with olive oil and place it over medium heat. Add all of the ingredients except the potatoes and chickpeas.

4. After a few minutes of cooking, pour in the chickpeas and stir while cooking for 5 minutes.

5. Slice the longer side of each potato down the middle leaving an opening. Stuff the opening with the chickpea mixture. Top it off with a tablespoon of the mixed yogurt.

6. Serve right away.

Zucchini-Basil Soup

Cooking Time: 45 minutes
Serves: 5

Ingredients:

- 2 pounds zucchini, trimmed and cut crosswise into thirds
- 3/4 cup chopped onion
- 2 garlic cloves, chopped
- 1/4 cup olive oil
- 4 cups water, divided
- 1/3 cup packed basil leaves

Directions:

1. Peel and julienne the skin from half of zucchini; toss with 1/2 teaspoon salt and drain in a sieve until wilted, at least 20 minutes. Coarsely chop remaining zucchini.

2. Cook onion and garlic in oil in a saucepan over medium-low heat, stirring occasionally, until onions are translucent. Add chopped zucchini and 1 teaspoon salt and cook, stirring occasionally.

3. Add 3 cups water and simmer with the lid ajar until tender. Pour the soup in a blender and purée soup with basil.

4. Bring remaining cup water to a boil in a small saucepan and blanch julienned zucchini. Drain.

5. Top soup with julienned zucchini. Season soup with salt and pepper and serve.

Rice with Summer Squash, Red Peppers, and Pepitas

Cooking Time: 30 minutes
Serves: 5

Ingredients:

- 1 1/2 cups long-grain white rice
- 3 cups low sodium chicken broth
- 1 medium yellow zucchini or crookneck squash, cut into 1/2-inch cubes
- 1 red bell pepper, cut into 1/2-inch cubes
- 1 medium green zucchini, cut into 1/2-inch cubes
- 3 tablespoons salty roasted pepitas (pumpkin seeds)

- 1 onion, finely chopped
- 3 tablespoons finely chopped cilantro or parsley
- 2 tablespoons olive oil
- Salt to taste
- Pepper to taste

Directions:

1. Pour olive oil on to large saucepan and turn up the heat to medium-high.

2. Add the chopped onions into the mixture and stir frequently while cooking until the onions soften.

3. Add the yellow zucchini (or squash), green zucchini and bell pepper to the saucepan. Sauté the vegetables until they start softening.

4. Add the white rice, sauté and keep stirring for about a minute. Sprinkle with salt and pepper while you're at it.

5. Turn up the heat even further and add broth. Wait for the broth to boil and then turn down the heat and cover. Let the rice cook for about 20 minutes. Check regularly to see if the rice is already tender.

6. Add more salt and pepper if necessary. Turn off the heat and stir in the pepitas and parsley.

Grilled Summer Vegetables with Harissa

Cooking Time: 15 minutes
Serves: 6

Ingredients:

- 1/2 cup plus 3 tablespoons extra-virgin olive oil
- 2 tablespoons fresh lemon juice
- 2 teaspoons harissa powder
- 1 red bell pepper, quartered, seeded
- 1 1 1/2-pound eggplant, trimmed, cut crosswise into 1/2-inch-thick rounds
- 4 medium zucchini, halved lengthwise, then crosswise
- 6 large green lettuce leaves
- 3 tablespoons chopped fresh cilantro
- Lemon wedges (for garnish)

Directions:

1. Whisk 1/2 cup oil, lemon juice, and harissa in medium bowl. Season with salt.

2. Prepare barbecue or broiler. Grill bell pepper quarters, skin side down, until skin is blackened all over (do not turn). Leave the peppers to cool.

3. Pour remaining 3 tablespoons of oil onto a baking sheet. Layer eggplant and zucchini on the baking sheet in a single layer; turn to coat. Sprinkle vegetables with salt and pepper. Grill until charred in spots and cooked through, turning occasionally.

4. When the peppers are cool to the touch, peel the skin off the peppers.

5. Line a serving platter with lettuce. Arrange vegetables on top of the lettuce in a circle, leaving the center empty. Cut bell pepper and zucchini into 1-inch pieces and stack in the middle. Drizzle some of harissa dressing over. Sprinkle with cilantro.

6. Serve with lemon wedges and extra dressing on the side.

Creamy Asian Chicken Noodle Soup

Cooking Time: 25 minutes
Serves: 4

Ingredients:

- 3 ounces of dried soba noodles
- 1/4 cup of chopped cilantro
- 1 tablespoon olive oil
- 1 yellow onion
- 1 cup of shelled edamame
- 1 tablespoons of fresh ginger
- 1 peeled & finely chopped carrot

- 1 clove of minced garlic
- 4 cups of chicken or vegetable stock
- 2 tablespoons of sodium-free soy sauce
- 1 pound of chopped skinless chicken breasts
- 1 cup of plain soy milk

Directions:

1. Fill pot with water and bring to a boil. Add noodles and cook until tender, drain and set aside.

2. Warm olive oil over medium heat in a medium saucepan. Sauté onions until translucent and soft, throw in ginger, garlic, and carrot and continue to sauté for about 1 minute. Add chicken and stir fry for about 3 minutes.

3. Add stock and soy sauce to saucepan and bring to a boil. Add edamame.

4. Reduce heat to low and let simmer until the chicken is fully cooked.

5. Place soba noodles and soy milk into the pot, cook until heated through but do not let it boil.

6. Remove from heat, stir in cilantro.

7. Serve immediately.

No-pasta Zucchini Lasagna

Cooking Time: 45 minutes
Serves: 6

Ingredients:

- 2 zucchinis, sliced 1/4-inch thick
- 4 tomatoes, sliced 1/4-inch thick
- 8 ounces mozzarella, shredded
- 2 onions, thinly sliced
- 1 tablespoon olive oil
- 1 sprig fresh basil (6-8 leaves), chopped
- Ground black pepper

Directions:

1. Prepare a medium casserole dish and grease it with olive oil.

2. Cover the bottom layer of the dish with sliced zucchini.

3. Add a layer of tomatoes and onions on top of the zucchini slices. Make sure the tomatoes and onions are spread out.

4. Cover the layer with some sliced basil and ground pepper followed by shredded mozzarella cheese. Make sure you set aside some basil and cheese for the top most layer.

5. Keep adding vegetables and cheese alternately until you run out.

6. Preheat your oven to 400°F and bake the lasagna for about 30 minutes. Wait 5 extra minutes so the lasagna can cool a bit and then serve immediately.

Bun-less Sliders

Cooking Time: 30
Serves: 2

Ingredients:

- 1 tablespoon olive oil
- 1/2 large onion, sliced very thin
- 1 small pat of butter
- 1/4 cup red wine
- 1 lb. extra-lean ground sirloin
- 4 slices Swiss cheese
- 2 cups grape tomatoes
- 4 iceberg lettuce leaves, washed and dried

Directions:

1. Preheat the oil in a nonstick skillet over medium to medium-low heat. Sauté the onions until translucent and soft. Stirring often. When translucent, add the butter and red wine to finish.

2. Make 4 medium sized, thin hamburger patties, using the ground sirloin. The patties will shrink when cooked.

3. Pan-fry them over medium-high heat in the same skillet used to cook the onions. Cook until the patties are fully cooked.

4. Plate the patties and onions with the lettuce leaves on the side. To serve, wrap a patty with some onions in the lettuce wrap to make a bun-less slider.

Rainbow Quinoa

Cooking Time: 15 minutes
Serves: 2

Ingredients:

- 1 cup Quinoa, cooked
- 1 cup frozen mixed vegetables such as peas, carrots, green beans, corn (or fresh)
- 1 tablespoon Lemon Juice
- 1 tablespoon Olive Oil
- Salt and Pepper to taste

Directions:

1. Defrost, or prep fresh, vegetables.

2. Mix quinoa, vegetables, lemon juice and olive oil thoroughly but gently.

3. Taste and season if necessary.

4. Serve chilled.

Spicy Baked Spinach and Cod

Cooking Time: 2 minutes
Serves: 2

Ingredients:

- Cooking oil spray
- 1 pound cod (or other fish) fillet
- 1 tablespoon olive oil
- 1 teaspoon spicy seasoning mix

Directions:

1. Preheat oven to 350 degrees.

2. Spray a casserole dish with cooking spray.

3. Wash the fish and pat dry.

4. Place the fish into the casserole dish and sprinkle with oil and seasoning mix.

5. Bake uncovered for 15 minutes or until fish flakes with fork.

6. Cut the fishes into 4 pieces.

7. Serve with rice.

Black Bean Salad Stuffed Sweet Potato

Cooking Time: 30 minutes
Serves: 2

Ingredients:

- Vegetable oil
- 2 large sweet potatoes, pricked
- 1 teaspoon salt
- 1/2 teaspoon freshly ground black pepper
- 1/4 cup fresh lime juice, plus wedges for garnish
- 1 tablespoon balsamic vinegar
- 1 tablespoon finely chopped garlic

- 1 can (15 ounces) black beans, rinsed and drained
- 1 cup halved cherry tomatoes
- 1/2 cup thinly sliced orange or red bell pepper
- 1/2 cup thinly sliced scallions
- 1/3 cup chopped fresh mint
- 4 cups baby arugula

Directions:

1. Rub the sweet potatoes with oil and season with salt and pepper; bake in an oven preheated to 375 F. Bake until tender, 15-25 minutes; let cool.

2. In a bowl, whisk juice, vinegar, garlic. Season with salt and pepper. Add beans, tomatoes, bell pepper, scallions and mint; toss.

3. To serve, slice a sweet potato in half, length wise and lightly mash the insides. Fill the sweet potato with the bean salad. Top with arugula, and garnish with lime wedges.

Chapter 6: DASH Diet Dessert Recipes

When you are on the DASH diet, you don't have to completely sacrifice your sweet tooth. The following recipes below are proof but try not to go too crazy on these and be mindful of your daily servings. Some of these recipes require coconut sugar which has a lower glycemic index than regular table sugar but it is still high in calories.

Mango Tapioca Rice Pudding

Cooking Time: 50 minutes
Serves: 2

Ingredients:

- 1 1/2 cup low fat milk
- 1/2 cup rice
- 1/4 cup tapioca
- 2 tablespoons coconut sugar
- 1/2 teaspoon vanilla extract
- 5 -6 drops almond extract
- A large pinch of cinnamon
- 1 mango, cubed
- 2 tablespoons almonds, toasted, chopped

Directions:

1. Place a heavy saucepan over medium heat. Add milk, tapioca and rice. Bring to a boil.

2. Lower heat and simmer until the rice is cooked.

3. Remove from heat. Add sugar, vanilla extract, and cinnamon. Stir in mangoes.

4. Serve either warm or cold sprinkled with almonds and more mango cubes on top.

Apple Pudding Cake

Cooking Time: 4 hours
Serves: 4

Ingredients:

- 4 cups apple, cored, finely chopped (do not peel)
- 4 cups coconut sugar
- 4 cups flour
- 4 eggs
- 2 cups vegetable oil
- 2 cups walnuts, chopped
- 2 teaspoons baking soda
- 1 teaspoon nutmeg, grated

Directions:

1. Mix together in a bowl, flour, nutmeg, and baking soda.

2. Add sugar, oil, eggs and vanilla to a large bowl and beat well.

3. Add the dry ingredient mixture and mix well. Add apples and fold.

4. Grease the inside of the slow cooker. Pour the batter into the cooker.

5. Slip a toothpick between the lid and the slow cooker to create a gap for excess steam. Do not open lid until cooking time is done.

6. Set the cooker on High for 3 1/2 to 4 hours. Slice when warm or cold and serve with a sweet sauce of your choice (optional).

Carrot Cupcakes with Lemon Icing

Preparation Time: 1 hour
Serves: 6

Ingredients:

Cupcake:

- 2 cups carrot pulp
- 1 cup dates, chopped
- 1 cup almonds
- 3/4 cup raisins
- 1 teaspoon cinnamon powder
- 1/2 teaspoon ginger, grated
- A pinch of nutmeg
- A pinch of salt

Icing:

- 6 dates, pitted, soaked for 1 hour and drained
- 1 cup cashews, soaked for 1 hour and drained
- 1 teaspoon lemon juice
- A splash of water
- A pinch of salt

Directions:

1. Place all of the cupcake ingredients in a food processor. Blend until mixed very well.

2. Fill up as many cupcake tins as you can with the cupcake mixture until you run out.

3. Wash and rinse your food processor and then add all of the icing ingredients. Blend them the same way you did with the cupcake ingredients.

4. Spread the frosting on each of the carrot cupcakes.

5. Serve cupcakes after an hour in the refrigerator.

Strawberry Fruit Sundae

Cooking Time: 10 minutes
Serves: 9

Ingredients:

- 1 1/2 cups strawberries, chopped
- 5 cups mixed fruits, chopped
- 8-10 waffle cones
- 1/2 cup jicama, shredded

Directions:

1. Puree the strawberries. Set them aside afterwards.

2. Mix together the rest of the fruits and scoop them on the waffle cones.

3. Top off the waffle cones with the pureed strawberries.

4. Add some shredded jicama and serve.

Holiday Pumpkin Pie

Cooking Time: 1 hour
Serves: 6

Ingredients:

- 2 cups ginger snaps, ground
- 32 ounces canned pumpkin
- 1 cup egg whites
- 1 cup coconut sugar
- 4 teaspoons pumpkin pie spice blend
- 2 cans (12 ounce each) evaporated skim milk

Directions:

1. Grease an ovenproof pie pan with cooking spray. Place the ground cookies in the pan. Spread well and press lightly.

2. In a large bowl, mix together the rest of the Ingredients. Pour over the cookies.

3. Bake in a preheated oven at 350 degree F for about 40-45 minutes.

4. Cool and refrigerate.

5. Slice into wedges and serve.

Pear Caramel Pudding

Cooking Time: 1 hour 15 minutes
Serves: 4

Ingredients:

- 1 cup all-purpose flour
- 1/3 cup granulated sugar
- 1 tablespoon flaxseed meal
- 1 teaspoon baking powder
- A pinch salt
- 1/2 teaspoon cinnamon powder
- 1/2 cup fat free milk
- 2 tablespoons canola oil

- 1/4 cup dried pears, snipped
- 1/2 cup water
- 1/2 cup pear nectar
- 6 tablespoons brown sugar
- 1 tablespoon butter
- Fat free vanilla yogurt (optional)
- Cooking spray

Directions:

1. In a bowl mix together flour, granulated sugar, flaxseed meal, and baking powder, cinnamon, and salt. Add milk and oil. Mix well. Add the dried pears.

2. Spray the inside of the slow cooker with cooking spray. Pour the batter into the cooker.

3. Meanwhile mix together water, pear nectar, brown sugar, and butter in a saucepan. Heat it and bring to a boil. Boil for 2 minutes.

4. Pour this sugar solution over the batter in the cooker. Cover. Set the cooker on low for 3 1/2 hours.

5. When done, switch off the cooker. Uncover and let it remain in the cooker for about 45 minutes.

6. Divide into bowls. Serve topped with yogurt.

Almond Figs

Cooking Time: 4 minutes
Serves: 3

Ingredients:

- 12 figs, halved
- 1 cup almonds blanched and toasted
- 2 tablespoons coconut butter
- 1/4 cup palm sugar

Directions:

1. Melt the coconut butter in a pot over high heat.

2. Stir in the figs and sugar and cook for 4 minutes.

3. Add the almonds and toss to coat

4. Serve in bowls.

Low-fat Chocolate Pudding

Cooking Time: 30 minutes
Serves: 2

Ingredients:

- 4 tablespoons cocoa powder
- 6 tablespoons cornstarch
- 4 cups nonfat milk
- 1 teaspoon vanilla
- 4 tablespoons coconut sugar or to taste
- 1/8 teaspoon salt
- 2/3 cup chocolate chips

Directions:

1. Whisk together the cocoa, milk, cornstarch, salt and coconut sugar in a bowl.

2. Transfer mixture to a large saucepan and place it over medium heat. Stir thoroughly until the mixture thickens.

3. Remove saucepan from heat source and add the vanilla and chocolate chips. Mix the pudding well until all chocolate chips are dissolved.

4. Prepare some serving bowls and pour the pudding into each of them. Cover with cling wrap and place them in the refrigerator. Serve chilled.

Sunday Chocolate Banana Cake

Cooking Time: 40 minutes
Serves: 4

Ingredients:

- 2 cups all-purpose flour
- 1/2 cup Splenda Brown Sugar Blend
- 1/4 cup unsweetened cocoa powder
- 1/2 teaspoon baking soda
- 1 large ripe banana, mashed (1/2 cup)
- 3/4 cup soy milk

- 1/4 cup canola oil
- 1 large egg
- 1 egg white
- 1 tablespoon lemon juice
- 1 teaspoon vanilla extract
- 1/2 cup semisweet dark chocolate chips

Directions:

1. Preheat oven to 350 F. Spray the bottom of an 11 x 7 inch brownie pan with nonstick spray.

2. In a large bowl combine together flour, brown sugar blend, cocoa, and baking soda.

3. Whisk together the bananas, soy milk, oil, egg, egg white, lemon juice, and vanilla in another bowl.

4. Make a hole in the middle of the flour mixture. Pour soy milk mixture and chocolate chips into the hole.

5. Stir the ingredients together with a wooden spoon, until combined. Tip the batter into the pan.

6. Bake about 25 minutes until the center of the cake springs back when pressed lightly with fingertips.

Summer Fresh Fruit Kebabs

Cooking Time: 15 minutes
Serves: 2

Ingredients:

- 6 ounces low-fat, sugar-free lemon yogurt
- 1 teaspoon fresh lime juice
- 1 teaspoon lime zest
- 4 pineapple chunks (about 1/2 inch each)
- 4 strawberries
- 1 kiwi, peeled and quartered
- 1/2 banana, cut into 4 1/2-inch chunks
- 4 red grapes
- 4 wooden skewers

Directions:

1. Whisk the yogurt, lime juice and lime zest in a small bowl.

2. Cover and chill in the refrigerator until needed.

3. Thread the one of each fruit onto a skewer. Repeat until all the fruit has been skewered.

4. Serve with the lemon lime dip.

Sunflower Energy Bars

Cooking Time: 30 minutes
Serves: 6

Ingredients:

- 1/3 pound walnuts
- 1/3 pound almonds
- 1/3 pound peanuts, chopped
- 1/4 sunflower seeds

- 2 cups honey
- 6 tablespoons sugar
- Water

Directions:

1. Fill up a baking sheet with all the nuts and seeds spread out. Place the baking sheet in a oven heated up to 350°F. Roast the nuts for 10 minutes.

2. Prepare a pan over medium heat and stir in the honey with the sugar until the mixture boils.

3. Transfer the nut mixture to the pan. Stir the nuts with the honey and let the mixture cook for 15 minutes more.

4. Set up a pan brushed with water and pour the mixture there. Spread the mixture evenly and set it aside.

5. Once cooled, turn the pan upside down, cut them into rectangular shapes and serve chilled.

Baked Brie Envelopes

Cooking Time: 25 minutes
Serves: 4

Ingredients:

- 1/2 cup fresh or frozen cranberries
- 1/2 medium orange, quartered
- 2 tablespoons sugar
- 1 cinnamon stick
- 1 sheet puff pastry dough, cut into 12 1/4-ounce squares
- 6 ounces Brie cheese, cut into 1/2-ounce cubes
- 2 tablespoons water
- 1 egg white

Directions:

1. Preheat oven to 425 F. Coat a small sauce pan with cooking spray and heat on medium-high heat

2. In the pan add the oranges, cranberries, sugar and cinnamon stick, and cook for about 10 minutes.

3. Stir constantly until cranberries are soft and the mixture starts to thicken.

4. Remove from heat and let it cool. Remove orange quarters and the cinnamon stick.

5. Roll out each square of puff pastry.

6. Onto each puff pastry square, place 1 teaspoon of cooled cranberry mixture and one cube of cheese.

7. Combine the water and egg white in a small bowl. Dab a small amount of the egg mixture onto the edges of the puff pastry.

8. Fold one corner of the pastry at a time around the cheese and cranberry mixture, like an envelope.

9. Baste the top of the pastry with the egg mixture.

10. Place the envelopes on a baking sheet and bake for 10 to 12 minutes or until golden brown.

Oven-baked Peaches with Greek Yogurt

Cooking Time: 20 minutes
Serves: 4

Ingredients:

- 6 peaches, halved and pitted
- 1/2 cup plain Greek yogurt, 2% fat
- 2 teaspoons honey
- 1 tablespoon unsalted butter, melted
- 1 tablespoon sugar
- 2 tablespoons hazelnuts, toasted and chopped

Directions:

1. Preheat your oven to 375°F and prepare a baking sheet with parchment. Place peaches with the cut side facing up on sheet.

2. Brush the peaches with butter and sprinkle with sugar.

3. Bake the peaches for about 15 minutes or until they get soft. Stop baking as soon as you see some juices run off.

4. Serve the peaches evenly among 4 bowls. Add 2 tablespoons of yogurt to each bowl and top off with hazelnuts and honey.

Cinnamon-baked Apple Slices

Cooking Time: 45 minutes
Serves: 4

Ingredients:

- 4 apples, cored, peeled, and thinly sliced
- 1/2 tablespoon ground cinnamon
- 1/4 teaspoon ground nutmeg
- 1/4 cup brown sugar
- 2 teaspoons freshly squeezed lemon juice (optional)

Directions:

1. Preheat your oven to 375°F.

2. Place the apples in a mixing bowl and combine them with the rest of the ingredients. Gently stir until apples are evenly coated.

3. Put the coated apples on a nonstick pan. Cover and place in the preheated oven.

4. Let the apple slices bake for 15 minutes. Before the timer is up, flip the apple slices and extend the timer to 30 more minutes.

5. Periodically check if the slices soften. Once soft, let the slices bake for a couple more minutes until the cinnamon coating thickens.

6. Serve immediately.

Almond and Sour Cherry Chocolate Bark

Cooking Time: 1 hour (for chocolate to set)
Serves: 24

Ingredients:

- 3/4 cup almonds, toasted and unsalted
- 12 ounces dark chocolate (60 percent to 70 percent cocoa)
- 1/2 teaspoon pure vanilla extract
- 1/3 cup dried tart cherries, roughly chopped

Directions:

1. Fill a medium saucepan with 2 inches of water; bring to a simmer over medium-low heat. Place a slightly larger heatproof bowl on top of the saucepan, making sure water doesn't touch bottom of bowl.

2. Place 10 ounces dark chocolate (60 percent to 70 percent cocoa) in bowl and stir, until smooth. Remove bowl from saucepan; add another 2 ounces dark chocolate and stir until smooth.

3. Stir in 1/2 teaspoon pure vanilla extract, toasted almonds and 1/3 cup dried tart cherries. Pour onto baking sheet and spread into an even layer about 1/4 inch thick.

4. Refrigerate until firm, 1 hour. Break into 24 pieces.

Chocolate-Mint "Nice Cream"

Cooking Time: 10 minutes
Serves: 4

Ingredients:

- 3 banana, frozen
- 3 tbsp Unsweetened Cocoa Powder
- 1/2 tsp peppermint extract

Directions:

1. Remove the sliced banana slices from the freezer and let stand for about 5 minutes.

2. Add the banana slices, cocoa, and peppermint extract to a food processor and pulse until the banana slices are finely chopped. Then puree until mixture resembles soft-serve ice cream.

Low Fat Semisweet Brownies

Cooking Time: 30 minutes
Serves: 16

Ingredients:

- 1 1/2 cups all-purpose flour
- 1 teaspoon baking powder
- 6 ounces semisweet chocolate
- 1 teaspoon vanilla extract
- 4 egg whites
- 2/3 cup granulated sugar
- 1/2 cup hot water
- 1/2 cup walnuts, chopped (optional)
-
-
-
-
-

Directions:

1. Melt chocolate in a large heatproof bowl set over simmering water; stir until smooth. Remove from heat and let cool slightly. Whisk in egg whites and followed by the vanilla.

2. In a separate bowl, mix together the sugar, flour, and baking powder; stir into chocolate batter until just combined. Fold in walnuts.

3. Spray an 8-inch cake pan with cooking spray. Pour batter into cake pan and bake in an oven preheated to 350 F for 20-25 minutes, or until edges pull away from pan.

4. Let the brownies cool slightly on a cooling rack, then slice and serve.

Sweet Tomato Cake

Cooking Time: 35 minutes
Serves: 6

Ingredients:

- 1-1/2 cups whole-wheat flour
- 1 cup tomatoes, blanched, deseeded, peeled and chopped
- 3/4 cup brown sugar
- 1/2 cup olive oil
- 2 tablespoons apple cider vinegar
- 1 teaspoon cinnamon, ground
- 1 teaspoon baking powder
- 1 teaspoon baking soda

Directions:

1. Prepare a mixture of baking powder, baking soda, flour, cinnamon and sugar in a bowl.

2. Place the remaining ingredients in another bowl.

3. Stir each individual bowl well and then stir in both mixtures together.

4. Pour the mixture into a greased round baking pan. Put the pan in an oven preheated to 350°F and let the cake bake for 35 minutes.

5. Serve after the cake cools down.

Decadent Oatmeal and Banana Cookies

Cooking Time: 25 minutes
Serves: 18

Ingredients:

- 4 ripe bananas
- 1/4 cup Palm sugar (coconut sugar)
- 1 1/2 cups old fashioned oats
- 1/2 cup dried cherries
- 1/2 cup dark chocolate chips
- Pinch of ground nutmeg
- 1/2 tsp cinnamon

Directions:

1. Preheat oven to 350 F and line a cookie sheet with parchment paper.

2. In medium bowl mash bananas. Add the rest of the ingredients and mix well.

3. Drop batter onto the prepared cookie sheet and bake for about 10 minutes, or until the bottoms of the cookie are golden brown.

Summer Fruit Salsa with Sweet Chips

Cooking Time: 3 hours 15 minutes
Serves: 4

Ingredients:

- 8 whole-wheat tortillas
- 1 tablespoon sugar
- 1/2 tablespoon cinnamon
- 3 cups diced fresh fruit, such as apples, oranges, kiwi, strawberries, grapes or other fresh fruit
- 2 tablespoons sugar-free jam, any flavor
- 1 tablespoon honey or agave nectar
- 2 tablespoons orange juice

Directions:

1. Preheat oven up to 350 F.

2. Cut each tortilla into 10 wedges and lay it flat in one layer onto a baking sheet. Spray cooking spray onto the tortilla pieces.

3. Combine sugar and cinnamon into a small bowl. Sprinkle evenly over the tortilla wedges.

4. Bake until the pieces are crisp.

5. Put on a cooling rack to let it cool.

6. Cut the fruit into cubes.

7. In a mixing bowl toss the fruit together gently.

8. Whisk together jam, honey and orange juice in another bowl.

9. Pour this over the diced fruit and mix gently again.

10. Cover the fruits with plastic wrap and refrigerate for 2 to 3 hours.

11. Serve with the cinnamon tortilla chips.

Chapter 7: DASH Diet Snack Recipes

Snacks can be healthy too! And because all of the recipes here serve more than one person, you better invite some friends or relatives and give them a taste to the DASH diet gateway!

All Dressed Crispy Potato Skins

Cooking Time: 1 hour 25 minutes
Serves: 2

Ingredients:

- 2 medium russet potatoes
- Butter-flavored cooking spray
- 1 tablespoon minced fresh rosemary
- 1/8 teaspoon freshly ground black pepper

Directions:

1. Preheat the oven to 375 F.

2. Wash the potatoes with fresh water.

3. Pierce the potatoes with a fork.

4. Bake the potatoes in an oven about 1 hour or until the skins are crisp.

5. Cut the hot potatoes in half and scoop out most of the flesh, leaving about 1/8 inch from the skin.

6. Use a butter-flavored cooking spray to spray the insides of each potato skin.

7. Press rosemary and pepper into the potato skin.

8. Bake the potato skins again for another 5 to 10 minutes.

9. Serve immediately.

The Best Grilled Pineapples

Cooking Time: 30 minutes
Serves: 2

Ingredients:

- 2 tablespoons dark honey
- 1 tablespoon olive oil
- 1 tablespoon fresh lime juice
- 1 teaspoon ground cinnamon
- 1/4 teaspoon ground cloves
- 1 firm, ripe pineapple

- 8 wooden skewers, soaked in water for 30 minutes, or metal skewers
- 1 tablespoon dark rum (optional)
- 1 tablespoon grated lime zest

Directions:

1. Use a charcoal grill or heat a gas grill or broiler.

2. Coat the grill rack or broiler pan with cooking spray. Set the cooking rack 4 to 6 inches from the heat source.

3. In a small bowl, combine the honey, olive oil, lime juice, cinnamon and cloves and whisk to make the marinade. Set aside.

4. Cut off the crown and the base of the pineapple.

5. Keep the pineapple upright and pare off the skin using a large, sharp knife, cutting downward just below the surface in long, vertical strips and leaving the small brown "eyes" on the fruit.

6. Lay the pineapple on its side. Remove all the eyes.

7. Put the peeled pineapple upright and cut it in half lengthwise.

8. Place each pineapple half cut-side down and cut it lengthwise into four long wedges; slice away the core.

9. Cut each wedge into three pieces. Thread the three pineapple pieces onto each skewer.

10. Brush each piece of pineapple with the marinade.

11. Grill or broil, turning once and basting once or twice with the remaining marinade, until tender and golden, about 5 minutes on each side.

12. Remove the pineapple from the skewers and place on a plate.

13. Finish with a brush of rum and a sprinkle of lime zest.

14. Serve hot or warm.

Mixed Fruit Skewers with Lemon Yogurt

Preparation Time: 10 minutes
Serves: 2

Ingredients:

- 4 pineapple pieces
- 4 strawberries
- 4 red grapes
- 4 medium banana pieces, freshly peeled
- 1 kiwi, peeled and quartered

- 6 ounces sugar-free and low fat lemon yogurt
- 1 teaspoon lime zest, grated
- 1 teaspoon lime juice

Directions:

1. Thread all the mixed fresh fruits on skewers and arrange them on a platter.

2. Whisk the lemon yogurt with the lime juice and lime zest. Serve chilled alongside your fruit skewers.

Quick Pickled Asparagus

Cooking Time: 12 hours
Serves: 4

Ingredients:

- 1 pound fresh asparagus, trimmed (about 3 cups)
- 1/4 cup pearl onions
- 1/4 cup white wine vinegar
- 1/4 cup cider vinegar
- 1 sprig fresh dill
- 1 cup water
- 2 whole cloves
- 3 cloves garlic, whole
- 8 whole black peppercorns
- 1/4 teaspoon red pepper flakes
- 6 whole coriander seeds

Directions:

1. Trim off the woody ends of the asparagus and cut the spears into lengths that will fit into the jars.

2. Put spears in colander and wash well and drain.

3. Trim the onions.

4. Mix all ingredients in air tight containers.

5. Preserve in the refrigerator up to 4 weeks.

Simple Shrimp Ceviche

Cooking Time: 3 hours 10 minutes
Serves: 2

Ingredients:

- 1/2 pound raw shrimp, cut in 1/4-inch pieces
- 2 lemons, zest and juice
- 2 limes, zest and juice
- 2 tablespoons olive oil
- 2 teaspoons cumin
- 1/2 cup diced red onion
- 1 cup diced tomato

- 2 tablespoons minced garlic
- 1 cup black beans, cooked
- 1/4 cup diced serrano chili pepper and seeds removed
- 1 cup diced cucumber, peeled and seeded
- 1/4 cup chopped cilantro

Directions:

1. Place shrimp in a shallow pan.

2. Zest one lemon and one lime. Squeeze the juice over the shrimp.

3. Freeze for at least 3 hours until shrimp is firm and white.

4. Combine the rest ingredients in separate bowl and set aside while shrimp is cold cooking.

5. Mix shrimp and the rest of the citrus juice with the remaining ingredients before serving.

6. Serve with baked tortilla chips.

Parmesan Stuffed Mushrooms

Cooking Time: 15 minutes
Serves: 20

Ingredients:

- 20 mushrooms, destemmed
- 2 cups basil, chopped
- 1-1/2 cups panko breadcrumbs
- 1/4 cup low-fat butter, melted
- 1/4 cup low-fat parmesan, grated
- 2 tablespoons parsley, chopped
- 2 tablespoons pumpkin seeds
- 2 teaspoons lemon juice
- 1 tablespoon garlic, minced
- 1 tablespoon olive oil
- A pinch of sea salt

Directions:

1. Combine the butter with panko breadcrumbs and parsley. Stir well and set the mixture aside.

2. Place the rest of the ingredients (except for the mushrooms) in your blender and pulse them really well.

3. Prepare a lined baking sheet and arrange the mushrooms on it. Top each mushroom with the mixture you made earlier.

4. Put the baking sheet in the oven preheated to 350°F. Let the mushrooms bake for 15 minutes.

5. Serve the mushrooms hot on a platter.

Southwestern Potato Skins

Cooking Time: 30 minutes
Serves: 6

Ingredients:

- 6 large baking potatoes
- 1 teaspoon olive oil
- 1 teaspoon chili powder
- 1/8 teaspoon Tabasco sauce
- 6 slices turkey bacon, cooked until crisp, chopped
- 1 medium tomato, diced
- 2 tablespoons sliced green onions
- 1/2 cup shredded cheddar cheese

Directions:

1. Preheat the oven to 450 F. Grease a baking sheet.

2. Scrub potatoes and prick each potato several times with a fork.

3. Microwave uncovered on high until tender, about 10 minutes. Let the potatoes cool on a wire rack.

4. Cut each potato in half lengthwise and scoop out the flesh, leaving about 1/4 inch of the flesh attached to the skin.

5. Whisk together the olive oil, chili powder and hot sauce in a small bowl.

6. Brush insides of the potato skins with the olive oil mixture.

7. Slice the potato skin into wedges and place onto the baking sheet.

8. Combine together the turkey bacon, tomato and onions in a small bowl.

9. Use this mixture to fill each potato skin and sprinkle each with cheese.

10. Bake to melt the cheese and until the potato skins are heated through, about 10 minutes.

11. Serve immediately.

Lemon and Garlic Broccoli

Cooking Time: 30 minutes
Serves: 4

Ingredients:

- 4 cups broccoli florets
- 1 cup water
- 1 tablespoon garlic, minced
- 1 teaspoon lemon zest
- 1 teaspoon olive oil
- Ground black pepper
- Salt

Directions:

1. In a small saucepan, bring 1 cup of water to a boil. Add the broccoli to the boiling water and cook for 2 to 3 minutes or until tender, being careful not to overcook. The broccoli should retain its bright-green color. Drain the water from the broccoli.

2. In a small sauté pan over medium-high heat, add the olive oil. Add the garlic and sauté for 30 seconds. Add the broccoli, lemon zest, salt, and pepper. Combine well and serve.

Soy Nut and Apricot Trail Mix

Cooking Time: 5 minutes
Serves: 10

Ingredients:

- 1 cup roasted soy nuts
- 1 cup roasted, shelled pistachios
- 1 cup pumpkin seeds
- 1 cup dried apricots, chopped
- 1 cup raisins

Directions:

1. Mix all ingredients in a bowl. Scoop into 1/4-cup portions, and place each portion in a zip-top snack bag.

Crispy Roasted Chickpeas

Cooking Time: 30 minutes
Serves: 2

Ingredients:

- 15 ounce canned chickpeas, drained and rinsed
- 2 teaspoons of your favorite spices and/or herbs
- 1/2 teaspoon olive oil
- 1/4 teaspoon salt

Directions:

1. Preheat your oven to 400°F.

2. Cover a rimmed baking sheet with a layer of paper towels. Spread the drained and rinsed chickpeas evenly on the sheet. Pat the chickpeas with paper towels until considerably dry.

3. Transfer the dried chickpeas into a bowl. Stir in the olive oil with a large spoon. Continue stirring while sprinkling the salt and your favorite herbs/spices.

4. Cover the baking sheet with a single layer of chickpeas.

5. Place the sheet in the oven and bake for 20 minutes. Before the timer ends, stir the chickpeas and then extend the timer to another 20 minutes.

6. Serve the roasted chickpeas as soon as they turn golden brown.

Smoked Trout Spread

Cooking Time: 5 minutes
Serves: 2

Ingredients:

- 1/4 pound smoked trout fillet, skinned and broken into pieces
- 1/2 cup 1 percent low-fat cottage cheese
- 1/4 cup coarsely chopped red onion
- 2 teaspoons fresh lemon juice
- 1 teaspoon hot pepper sauce
- 1/2 teaspoon Worcestershire sauce
- 1 celery stalk, diced

Directions:

1. Combine the trout, cottage cheese, red onion, lemon juice, hot pepper sauce and Worcestershire sauce in a blender or food processor.

2. Process until smooth, stopping to scrape down the sides of the bowl as needed.

3. Fold in the diced celery.

4. Keep in an air-tight container in the refrigerator.

Zucchini Tomato Sauce

Cooking Time: 50 minutes
Serves: 6

Ingredients:

- 2 tablespoons olive oil
- 2 small onions, chopped
- 3 cloves garlic, chopped
- 1-1/4 cup zucchini, sliced
- 1 tablespoon oregano, dried
- 1 tablespoon basil, dried
- 1 can tomato sauce
- 1 can tomato paste
- 2 medium tomatoes, chopped
- 1 cup water

Directions:

1. In a medium skillet, heat oil and sauté onions, garlic, and zucchini in oil until fragrant, about 5 minutes.

2. Add remaining ingredients and simmer covered for 45 minutes.

3. Serve over pasta.

Lemon White Bean Dip

Preparation Time: 10 minutes
Serves: 8

Ingredients:

- 15 ounces canned white beans, drained
- 8 garlic cloves
- 2 tablespoons lemon juice
- 2 tablespoons olive oil

Directions:

1. Place the garlic cloves in an oven preheated to 350°F. Let the garlic roast for about 40 minutes and wait for them to cool.

2. Blend all ingredients including the roasted garlic in your food processor.

3. Serve immediately with pita wedges or bread.

Easy Spinach Mango Salsa

Cooking Time: 40 minutes
Serves: 4

Ingredients:

- 2 blood or navel oranges or tangelos, peeled & sectioned (remove seeds)
- 5 cups fresh spinach leaves, washed & stems removed
- 3 cups red leaf lettuce, torn into bite-size pieces
- 1 cup peeled, diced ripe mangoes
- 1 small red onion, sliced razor thin, separated into rings
- 1 cup toasted pecan halves
- Add chicken if desired

Balsamic Orange Spice Dressing:

- 1/3 cup balsamic vinegar
- 1 tablespoon Tortuga Light Rum (add more if desired)
- 1/3 cup orange juice (from salad oranges, adding more if necessary)
- 2/3 cup olive oil
- 1/4 cup teaspoon ground allspice
- 1 teaspoon Tortuga Hellfire Hot Pepper Sauce

Directions:

1. Peel and remove the white membrane & fiber from the citrus fruits. Cut the citrus fruits into bite sized pieces. Chill until ready to use.

2. Combine all ingredients (balsamic vinegar, Tortuga Light Rum, orange juice, olive oil, ground allspice, Tortuga Hellfire Hot Pepper Sauce) into a jar.

3. Cover jar and shake well.

4. Place in a refrigerator for several hours, or until well-chilled.

5. Combine the spinach, lettuce, oranges, onion, mango & pecans in a large bowl.

6. Mix desired amount of dressing and toss lightly.

4 Week Meal Plan
Week 1

	Monday	Tuesday	Wednesday	Thursday	Friday	Saturday	Sunday
Breakfast	Egg Toast with Avocado Spread	Cinnamon Pumpkin Waffles	Summer Fruit Smoothie	Special Tofu Scramble	Herbed Wild Mushroom Oatmeal	Cranberry Green Tea Smoothie	Strawberry-Orange Low-fat Yogurt Milkshake
Lunch	Fresh Mediterranean Quinoa Salad	Shrimp Marinated in Lime Juice and Dijon Mustard	Grilled Portobello Mushroom Burgers	Seafood Chowder	Thousand Island Tuna Melt	Grilled Eggplant with Toasted Spices	Delicious Coconut Quinoa Curry
Snack	Apple Pudding Cake	Crispy Roasted Chickpeas	Southwestern Potato Skins	Almond Figs	Easy Spinach Mango Salsa	Soy Nut and Apricot Trail Mix	Chocolate-Mint "Nice Cream"
Dinner	Baked Almond Chicken	Mustard-coated Pork Tenderloin	DASH Beef Stew	Spinach-Stuffed Turkey Burger Patties	Ricotta and Pomegranate Bruschetta	Mexican Bake	Hearty Jambalaya

Week 2

	Monday	Tuesday	Wednesday	Thursday	Friday	Saturday	Sunday
Breakfast	Special Egg Scramble	Whole Wheat-Oat Pancakes with Blueberry Compote	Slim-down Smoothie	Cinnamon Breakfast Quinoa	DASH Oatmeal Special	Non-fat Strawberry Banana Smoothie	Fresh Fruit Crunch
Lunch	Grilled Chicken Salad	Spinach Artichoke Angel Hair Pasta with Shrimp	Shiitake Mushroom Chili	Roasted Garlic & Tomato Soup	Honey Whole Wheat Bread	Fresh Veggie Quesadillas	Fresh Tomato Crostini
Snack	Cinnamon-baked Apple Slice	The Best Grilled Pineapples	Sweet Tomato Cake	Quick Pickled Asparagus	Lemon and Garlic Broccoli	Summer Fruit Salsa with Sweet Chips	Chocolate Banana Cake
Dinner	Creamy Asian Chicken Noodle Soup	Rainbow Quinoa	Autumn Pork Chops	Spicy Baked Spinach and Cod	Rotisserie Chicken Salad with Creaming Tarragon Dressing	Bun-less Sliders	Chicken Tikka Masala

Week 3

	Monday	Tuesday	Wednesday	Thursday	Friday	Saturday	Sunday
Breakfast	Creamy Fruity French Toast	Whole Wheat Walnut Pancakes	Detox Smoothie	Whole Wheat Sweet Potato Cakes	Apple and Raisin Oatmeal	Layered Mango Green Smoothie	Spicy Strawberry Smoothie
Lunch	Cashew Chicken Salad	Chipotle Spiced Shrimp	Basil Pesto Stuffed Mushrooms	Vegan Lentil Soup	Cheesy Zucchini Pizza	**Summer Vegetable Rice**	Roasted Butternut Squash
Snack	Holiday Pumpkin Pie	Simple Shrimp Ceviche	Mango Tapioca Rice Pudding	Parmesan Stuffed Mushrooms	Low-fat Chocolate Pudding	All Dressed Crispy Potato Skins	Almond and Sour Cherry Chocolate Bark
Dinner	Ginger-Marinated Grilled Portobello Mushrooms	Healthy Brown Rice Pilaf	Beef and Bean Jalapeño Chili	Spicy Baked Spinach and Cod	Chicken Veggie Wrap	No-pasta Zucchini Lasagna	Grilled Chicken and Ratatouille

Week 4

	Monday	Tuesday	Wednesday	Thursday	Friday	Saturday	Sunday
Breakfast	Irish Brown Soda Bread	Banana Oat Pancakes With Spiced Syrup	Green Smoothie	Zucchini-Lemon Muffins	Bread Pudding	Banana Peanut Butter Shake	Apple Cinnamon Crisp
Lunch	Pork and Peach Salad with Walnuts	Easy Coconut Shrimp	Marinated Portobello Mushrooms with Provolone	Pork and Shiso Soup	Whole Wheat Pizza Margherita	Zucchini Pad Thai	DASH Chicken Noodles Soup
Snack	Oven-baked Peaches with Greek Yogurt	Mixed Fruit Skewers with Lemon Yogur	Carrot Cupcakes with Lemon Icing	Strawberry Fruit Sundae	Pear Caramel Pudding	Zucchini Tomato Sauce	Almond and Sour Cherry Chocolate Bark
Dinner	Ground Turkey Meatloaf	Roasted Vegetable Farro Salad	Simple and Delicious Beef Stew	Middle Eastern Hummus Wrap	Spicy Roasted Broccoli	Zucchini-Basil Soup	Roasted Brussels Sprouts and Chicken Dinner

Conclusion

Congratulations for reaching the end of the book! I hope you realize the vast possibilities in getting into the DASH diet even after you acknowledge all of the salty foods you have to avoid. Giving up might not be easy at first but you have to remember the central focus of this diet. Try not to focus on the weight loss component and keep the main benefit in your mind at all times. You read this book because you wanted to know more about hypertension and how to stop it and now you should know the dangers of letting hypertension get the better of you. It is never too late to change your eating habits and you don't have to do anything seriously drastic compared to other diets.

Stay focused on your objectives and use the recipes as a guide on what ingredients to buy from your local store. Doing this will lead you to that unexpectedly pleasant feeling of shedding off pounds. Don't stop there either because you want to stick with these habits and make sure you drive those hypertension demons away.

It all starts with getting to know the most important ingredients and gradually changing your supermarket shopping habits. The meal plan is full of ideas for your first month and nothing should stop you from coming up with your own. Set and try to achieve your health goals and you will experience the very best of what the DASH Diet has to offer.

Good luck!

Made in the USA
Lexington, KY
17 March 2018